GSAS

PUBLICATIONS SERIES

TYPOLOGIES: GSAS DESIGN GUIDELINES

v2.0 - 2013

COMMERCIAL
CORE + SHELL
RESIDENTIAL
EDUCATION
MOSQUES
HOTELS
LIGHT INDUSTRY
SPORTS

Dr. Yousef Al Horr
Founding Chairman

A MESSAGE FROM

DR. YOUSEF MOHAMED AL HORR,
FOUNDING CHAIRMAN

GORD has come a long way since pioneering the Global Sustainability Assessment System (GSAS), formerly known as (QSAS), the Middle East's first integrated and performance-based green building assessment rating system in 2009.

Our mission to encourage the development and implementation of sustainability principles and imperatives stems from the sustainable goals outlined in His Highness, The Emir Sheikh Hamad bin Khalifa Al-Thani's Qatar National Vision 2030, which aims to achieve sustainable economic development and environmental leadership.

GSAS draws from top-tier global sustainability systems and adds new facets and dimensions to the current practices in assessing the sustainability of the built environment. Modelled on best practices from the most established global rating schemes including, but not limited to, BREEAM (United Kingdom), LEED (United States), GREEN GLOBES (Canada), CEPAS (Hong Kong), CASBEE (Japan), and the International SBTOOL, GSAS has grown into a pan-regional system offering a comprehensive framework, and equally flexible to incorporate the specific needs of the local context of different regions. In Qatar, GSAS is currently the only rating scheme to be acknowledged by Qatar Construction Specifications (QCS 2010).

Primary goals of GSAS include creating a better living environment, minimizing resource consumption and reducing environmental degradation due to the fast pace of urbanization taking place in this era. Such objectives, coupled with the increasing evidence of climate change effects on a global level, have contributed strongly to the unprecedented pace of adaptation to sustainability practices not only in the developed countries, but also in developing countries at a pace that is unexpected.

GSAS Version 2.0 has become the most comprehensive system, to date, that addresses the built environment from a macro level to a micro level targeting a wide range of building typologies. The new system will have design assessments for all typologies integrated into one comprehensive manual. The manual provides recommendations and guidelines for the effective implementation of the sustainability goals of each criterion. As more research is carried out on the rating system, the manuals will be further developed to keep users informed on updates within the constantly evolving GSAS rating systems.

I would like to acknowledge the efforts and contributions from the State of Qatar, all our members, and international partners-especially the TC Chan Center for Building Simulation and Energy Studies at the University of Pennsylvania and the Center's associated consultants who helped establish the system and take it into new dimensions. Last but not least, the continuous support from Qatari Diar Real estate Investment Company is highly appreciated, and without its support, GSAS would not be able to achieve what it has achieved in such a short time.

DESIGN GUIDELINES

TABLE OF CONTENTS

DESIGN GUIDELINES

ACKNOWLEDGMENTS

GLOBAL SUSTAINABILITY ASSESSMENT SYSTEM (GSAS)
ACKNOWLEDGEMENT PREFACE

THIS PROJECT WAS INITIATED, COMMISSIONED AND LED BY

Dr. Yousef M Alhorr,
Founder and Chairman,
Gulf Organisation for Research and Development

PRINCIPAL PROJECT DEVELOPER AND DIRECTOR

Dr. Ali Malkawi
Professor of Architecture and Chairman of the Graduate Group, University of Pennsylvania,
Founder and Director, T.C. Chan Center for Building Simulation and Energy Studies

SPECIAL ACKNOWLEDGMENT

HE. Mr. Ghanim Bin Saad Al Saad
Chairman and Managing Director - Barwa

Eng. Mohammed Alhedfa,
GCEO, Qatari Diar Realestate Investment Company - State of Qatar

Dr. Mohammed Saif Alkuwari,
Under Secretary of Ministry Of Environment - State of Qatar

TECHNICAL LEAD AND DEVELOPER

Dr. Godfried Augenbroe,
Chair of Building Technology, Doctoral Program,
Professor, College of Architecture

DIRECTOR RESEARCH & DEVELOPMENT

Dr. Esam Elsarrag,
Gulf Organisation For Research & Development

ACKNOWLEDGMENTS

GULF ORGANISATION FOR RESEARCH AND DEVELOPMENT - QATAR

Team Leads

Hassan Satti   Ali Diyab Elsheikh

Technical Team

Abdulrahim Alsayed   Ronaldo Dalistan   Lakshmi Suryan
Omeima Khidir   Hassan El Aref   Mutasim Salim
Alek Zivkovic

Logistic and Support Lead

Salah Al Ayoubi

Logistic and Support Team

Ibrahim Aburhaiem   Murad Ali Naz   Khalid Radi
Mohammed Imran   Mohammed Dad   Mohammed Elsheikh
Melody Manaman   Gloria Pineda

TC CHAN CENTER – UNIVERSITY OF PENNSYLVANIA - USA

Team Leads

Chau Nguyen, Project Manager   Yun Kyu Yi, Assistant Project Manager

Research and Development

Yasmin Bhombal   Bin Yan   Khaled Tarabieh
Sean Williams   Aroussiak Gabrielian Lily Trinh Ciammaichella
Joseph Hoepp   Charles Nawoj   Alex Muller
Sarah Savage   Kristen Sterner   Alexander Waegel
Kristen Albee   Niketa Laheri   Jeremy Krotz
Rebecca Lederer   Ethan Leatherbarrow Noelle Tay
Erin Lauer   Yoon Soo Lee

Web Development

Brandon Krakowsky   Marcus Pierce   Sunil Kamat
Sibasish Acharya   Ruchir Jha

ACKNOWLEDGMENTS

AFFILIATED RESEARCH INSTITUTIONS
GEORGIA INSTITUTE OF TECHNOLOGY - USA

Research and Development Team

Sang Hoon Lee	Yeonsook Heo	Fei Zhao
Reen Foley		

QATARI GOVERNMENT AND SEMI-GOVERNMENT SECTOR

Barwa Real Estate Company (BARWA)
Lusail Real Estate Development Company (LUSAIL)
Ministry of Endowment and Islamic Affairs (AWQAF)
Ministry of Environment (MOE)
Ministry of Interior – Internal Security Forces (ISF)
Ministry of Municipal Affairs and Urban Planning (MMUP)
Private Engineering Office – Amiri Diwan (PEO)
Public Works Authority (ASHGHAL)
Qatari Diar Real Estate Investment Company (QD)
Qatar General Electricity and Water (KAHRAMA)
Qatar Museums Authority (QMA)
Qatar Olympic Committee (QOC)
Qatar Petroleum (QP)
Qatar Science and Technology Park – Qatar Foundation (QSTP)
Qatar University (QU)

QATARI PRIVATE SECTOR

Arab Engineering Bureau (AEB)   HOARE LEA Qatar (HLQ)
KEO International Consultants   Energy City Qatar (Energy City)

REGIONAL PROFESSIONAL ORGANISATIONS

State of Kuwait – Green Building Committee – National Codes Committee
Kingdom of Jordan – Jordanian Engineers Association
Republic of Sudan - University of Khartoum

ACKNOWLEDGMENTS

INTERNATIONAL EXPERT REVIEWERS AND CONSULTANTS

- **Dick Van Dijk, PhD [Netherlands]**

  Member of ISO TC163 Energy Standardization Committee, TNO, Institute of Applied Physics.

- **Frank Matero, PhD [US]**

  Professor of Architecture and Historic Preservation, University of Pennsylvania.

- **Greg Foliente, PhD [Australia]**

  Principal Research Scientist, CSIRO (Commonwealth Scientific and Industrial Research Organisation) Sustainable Ecosystems.

- **John Hogan, PE, AIA [US]**

  City of Seattle Department of Planning and Development, Member of ASHRAE.

- **Laurie Olin, RLA, ALSA [US]**

  Partner, OLIN Studio.

- **Mark Standen [UK]**

  Building Research Establishment Environmental Assessment Method (BREEAM) Technical work.

- **Matthew Bacon, PhD, RIBA, FRSA [UK]**

  Professor, University Salford - Faculty Built Environment and Business Informatics; Chief Executive, Conclude Consultancy Limited; and Partner, Eleven Informatics LLP.

- **Matt Dolf [Canada]**

  Assistant Director, AISTS (International Academy of Sports Science and Technology).

- **Matthew Janssen [Australia]**

  Director of Construction and Infrastructure and Environmental Management Services Business Units (KMH Environmental); formerly the Sustainability Program Manager for Skanska.

- **Muscoe Martin, AIA [US]**

  Director, Sustainable Buildings Industries Council (SBIC), USGBC board member.

ACKNOWLEDGMENTS

- **Nils Larsson [Canada]**

  Executive Director of the International Initiative for a Sustainable Built Environment (iiSBE).

- **Raymond Cole, PhD [Canada]**

  Director, School of Architecture and Landscape Architecture, University of British Columbia.

- **Skip Graffam, PhD, RLA, ASLA [US]**

  Partner, Director of Research, OLIN Studio.

- **Sue Riddlestone [UK]**

  Executive Director & Co-Founder of BioRegional, Co-Director of One Planet and M.D. of BioRegional MiniMills Ltd.

DESIGN GUIDELINES

PREFACE

The primary objective of Global Sustainability Assessment System (GSAS) is to create a sustainable built environment that minimizes ecological impact while addressing the specific regional needs.

The GSAS manuals and documents developed to date include the following:

- Construction: GSAS Guidelines v2.0
- Construction: GSAS Assessment v2.0
- Districts: GSAS Guidelines v2.0
- Districts: GSAS Assessment v2.0
- GSAS Energy Application v2.0
- GSAS Training Manual: Commercial & Residential – Part I v2.1
- GSAS Training Manual: Commercial & Residential – Part II v2.1
- GSAS Technical Guide v2.0
- Health Care: GSAS Design Guidelines v2.0
- Health Care: GSAS Design Assessment v2.0
- Parks: GSAS Guidelines v2.0
- Parks: GSAS Assessment v2.0
- Railways: GSAS Design Guidelines v2.0
- Railways: GSAS Design Assessment v2.0
- RFP Preparation: GSAS All Typologies v2.0
- Sports: GSAS Design Guidelines v2.0
- Sports: GSAS Design Assessment v2.0
- Typologies: GSAS Design Guidelines v2.0
- Typologies: GSAS Design Assessment v2.0
- Typologies: GSAS Operations Guidelines v2.0
- Typologies: GSAS Operations Assessment v2.0

DESIGN GUIDELINES

OBJECTIVE

The objective of the Design Guidelines is to provide recommendations and guidance for the effective implementation of the sustainable goals of each criterion within the Design Assessment System. The guidelines are intended to supplement the Design Assessment in facilitating the design of a sustainable built environment that minimizes ecological impact while addressing the regional needs and environmental conditions specific to Qatar.

Each criterion in the Design Assessment has an associated guideline that provides designers with descriptive information for consideration to help attain the specific credit. These suggestions are in the form of recommended methods, strategies, and technologies. Individual projects should consider and assess the potential advantages and benefits of the recommended design guidelines in relationship to the specific goals, requirements, and conditions of the project.

The Design Guidelines take advantage of the combined best practices and recommendations provided by multiple established rating systems, while integrating these best practices with the specific ecological conditions, sustainable objectives, and goals of Qatar. The guidelines are not meant to provide specific or explicit instruction on how to design a sustainable built environment, but rather to provide guidance and recommendations on how to approach the design issues within each criterion. Furthermore, these guidelines are by no means inclusive of all possible recommendations. Thus, all projects are ultimately expected to perform the research and analysis necessary for their specific conditions and goals in order to meet the sustainability requirements of the Design Assessment System.

Refer to the GSAS Design Assessment, for a detailed description of the scope for each building typology. For a summary of the list of criteria that are included for each building typology, refer to the CRITERIA SUMMARY section.

CRITERIA SUMMARY

The following chart summarizes the GSAS Design criteria applicable to each typology.

Legend	
Criteria Included in Scope of Rating System	●
Inherited (For Residential Type Only)	I
Not Currently Scored	N/A
Out of Scope/Not Rated	

		Commercial	Core + Shell	Single Residential	Group Residential	Education	Mosques	Hotels	Light Industry	Sports
UC	**Urban Connectivity**									
UC.1	Proximity to Infrastructure	●	●		●	●	●	●	●	●
UC.2	Load on Local Traffic Conditions	●	●	I	I	●	●	●	●	●
UC.3	Public Transportation	●	●	●	●	●	●	●	●	●
UC.4	Private Transportation	●				●		●	●	●
UC.5	Sewer & Waterway Contamination	●	●			●	●	●	●	●
UC.6	Acoustic Conditions	●	●	●	●	●	●	●		
UC.7	Proximity to Amenities	●	●	●	●	●		●	●	
UC.8	Accessibility	●	●			●	●			●

Table 1   Design Rating System Scope, Part I

		Commercial	Core + Shell	Single Residential	Group Residential	Education	Mosques	Hotels	Light Industry	Sports
S	**Site**									
S.1	Land Preservation	•	•	|	|	•	•	•	•	•
S.2	Water Body Preservation	•	•	|	|	•	•	•	•	•
S.3	Habitat Preservation	•	•	|	|	•	•	•	•	•
S.4	Vegetation	•	•	|	•	•	•	•	•	•
S.5	Desertification	•	•	|	•	•	•	•	•	
S.6	Rainwater Runoff	•	•		•	•	•	•	•	•
S.7	Heat Island Effect	•	•		•	•	•	•	•	•
S.8	Adverse Wind Conditions	•	•		•	•	•	•		
S.9	Noise Pollution	•	•			•	•	•	•	•
S.10	Light Pollution	•	•			•		•	•	•
S.11	Shading of Adjacent Properties	•	•		•	•	•	•	•	•
S.12	Parking Footprint	•	•		•	•	•	•	•	•
S.13	Shading	•	•	|	•	•	•	•	•	•
S.14	Illumination	•	•		•	•	•	•	•	•
S.15	Pathways	•	•		•	•	•	•	•	•
S.16	Mixed Use	•	•							
E	**Energy**									
E.1	Energy Demand Performance	•	•	•	•	•	•	•	•	•
E.2	Energy Delivery Performance	•	•	•	•	•	•	•	•	•
E.3	Fossil Fuel Conservation	•	•	•	•	•	•	•	•	•
E.4	CO_2 Emissions	•	•	•	•	•	•	•	•	•
E.5	NO_x, SO_x, & Particulate Matter	•	•	•	•	•	•	•	•	•
W	**Water**									
W.1	Water Consumption	•	•	•	•	•	•	•	•	•
M	**Materials**									
M.1	Regional Materials	•	•	•	•	•	•	•	•	•
M.2	Responsible Sourcing of Materials	•	•	•	•	•	•	•	•	•
M.3	Recycled Materials	•	•	•	•	•	•	•	•	•
M.4	Materials Reuse	•	•	•	•	•	•	•	•	•
M.5	Structure Reuse	•	•			•		•	•	
M.6	Design for Disassembly	•	•			•		•	•	•
M.7	Life Cycle Assessment (LCA)	N/A	N/A	N/A	N/A	N/A	N/A	N/A	N/A	N/A

Table 2   Design Rating System Scope, Part II

		Commercial	Core + Shell	Single Residential	Group Residential	Education	Mosques	Hotels	Light Industry	Sports
IE	**Indoor Environment**									
IE.1	Thermal Comfort	●	●			●	●	●	●	●
IE.2	Natural Ventilation	●	●	●	●	●	●	●	●	●
IE.3	Mechanical Ventilation	●	●			●	●	●	●	
IE.4	Illumination Levels	●	●			●	●	●	●	●
IE.5	Daylight	●	●	●	●	●	●	●	●	●
IE.6	Glare Control	●	●			●			●	●
IE.7	Views	●	●			●				●
IE.8	Acoustic Quality	●	●	●	●	●	●	●	●	●
IE.9	Low-Emitting Materials	●	●	●	●	●	●	●	●	●
IE.10	Indoor Chemical & Pollutant Source Control	●	●			●		●	●	●
CE	**Cultural & Economic Value**									
CE.1	Heritage & Cultural Identity	●	●	●	●	●	●	●		●
CE.2	Support of National Economy	●	●	●	●	●	●	●	●	●
MO	**Management & Operations**									
MO.1	Commissioning Plan	●	●		●	●	●	●	●	●
MO.2	Organic Waste Management	●	●		●	●		●	●	●
MO.3	Recycling Management	●	●		●	●	●	●	●	●
MO.4	Leak Detection	●	●			●	●	●	●	●
MO.5	Energy & Water Use Sub-metering	●	●			●		●	●	●
MO.6	Automated Control System	●	●			●	●	●	●	●
MO.7	Hospitality Management Plan					●		●		●
MO.8	Sustainability Education & Awareness Plan									●
MO.9	Building Legacy									●

Table 3   Design Rating System Scope, Part III

URBAN CONNECTIVITY [UC]

The design of the proposed project has a direct impact on adjacent buildings, properties, neighborhoods, and the larger urban community. Sustainable urban practices can improve and further the development of existing neighborhoods and communities, as well as minimize impacts on the surrounding environment including climate change, fossil fuel depletion, water depletion and pollution, air pollution, land use and contamination, and human comfort and health. Consider the following factors when designing the project: the proximity to existing infrastructure; the additional load on local traffic conditions; the effectiveness of public transportation to and from the site; the effect of the project on sewers and waterways; the noise levels surrounding the site; the connectivity between the new development and existing transportation infrastructure, as well as other buildings and amenities; and the accessibility of the site for pedestrians and bicyclists.

Criteria in this category include:

UC.1	Proximity to Infrastructure
UC.2	Load on Local Traffic Conditions
UC.3	Public Transportation
UC.4	Private Transportation
UC.5	Sewer & Waterway Contamination
UC.6	Acoustic Conditions
UC.7	Proximity to Amenities
UC.8	Accessibility

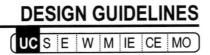
[UC.1] Proximity to Infrastructure

SCOPE

COMMERCIAL | CORE + SHELL | RESIDENTIAL | EDUCATION | MOSQUES | HOTELS | LIGHT INDUSTRY | SPORTS

- Rated for GROUP RESIDENTIAL

DESCRIPTION

Minimize the amount of new infrastructure construction by selecting a site near available connections to existing infrastructure or seeking on-site "off-grid" solutions.

GUIDELINES

This criterion measures the amount of additional infrastructure that must be constructed to serve a new building project. The criterion does not measure the actual impact of individual infrastructure projects, as such a projection would be too complex and subjective to measure for every project. Instead, the criterion treats each type of infrastructure equally and simply determines what percentage of the infrastructure needed by the project is already available on-site, or at an adjoining site, in order to reduce the construction of new infrastructure. Therefore, to improve the criterion score, the project must meet its infrastructural needs with services that are already available on-site or on an adjoining site. The project can also eliminate the need for a given type of infrastructure, either by removing the need for it from the design or by meeting that need in an alternative fashion.

If a project requires natural gas connections, but no connections exist nearby or on-site, then consider using electricity to meet the project's needs instead, providing that the existing electrical infrastructure can handle the additional load. By using existing infrastructure or by eliminating the need for new infrastructure connections, the project will reduce the overall costs and environmental impacts associated with infrastructure construction. Additionally, if the existing infrastructure is insufficient to meet the project's needs, efficiency improvements may eliminate the need for new infrastructure development.

SPORTS

Sports facilities often consume larger amounts of energy in bursts due to a larger amount of users during peak events. If a sports facility requires additional electric transmission capacity to meet the demands of peak events, avoid additional construction by using on-site generators to meet the extra demand if events occur infrequently.

FURTHER RESOURCES

Publications:

1. Sustainable Infrastructure Action Plan 2009-2011; World Bank Group, 2008.

2. Sustainable Infrastructure Initiative: Interdepartmental Planning for Better Capitol Projects; City of Seattle Department of Planning and Development, 2009.

3. Indicators and Framework for Assessing Sustainable Infrastructure; Shovini Dasgupta and Edwin K.L. Tam. Canadian Journal of Civil Engineering, Vol. 32, 30-44. 2005.

4. Barriers to Sustainable Infrastructure in Urban Regeneration; D. V. L. Hunt and C. D. F. Rogers. Proceedings of the ICE - Engineering Sustainability, Vol 158, Issue 2, 67-81. 2005.

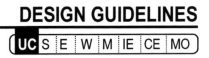
[UC.2] Load on Local Traffic Conditions

SCOPE

COMMERCIAL | CORE + SHELL | RESIDENTIAL | EDUCATION | MOSQUES | HOTELS | LIGHT INDUSTRY | SPORTS

• Inherited for SINGLE and GROUP RESIDENTIAL

DESCRIPTION

Minimize the impact on the local traffic conditions.

GUIDELINES

In order to reduce excessive vehicle emissions and roadway congestion, employ design strategies to decrease the number of vehicles on adjacent roads. Reduce the load on local infrastructure by providing easy and convenient access to public transportation, minimizing parking needs, limiting the use of private transportation, and designating areas on the site for the loading and unloading of guests and goods. Improve pedestrian and bicycle access to further reduce the transportation load. Encourage bicycle transit by providing designated road space for cyclists with care given to road junctions and other zones that may cause conflicts for various vehicles. Pedestrian travel can be aided with safe, well-marked sidewalks and path systems. See [UC.3] Public Transportation and [UC.4] Private Transportation for detailed recommendations on these modes of transportation.

Conduct studies to calculate and analyze the impact of future traffic loads on the existing surrounding infrastructure. Dynamic traffic simulations may be used to determine the projected loads on local roads. Consider loads on existing transportation networks when planning the placement of access roads and parking lot entries and exits on the site. Ensure that existing roads and intersections have the capacity to handle future traffic loads from the new development.

Provide an appropriate amount of parking on the site and reduce parking needs by providing alternative modes of transportation. Select a site in close proximity to existing public transport systems to promote the use of public transportation, thereby reducing the need for private cars and minimizing traffic congestion.

The project should include parking facilities and loading docks for the regular loading and unloading of goods as appropriate for the specific project. Providing zones for loading and unloading for both guests and goods on the site will relieve traffic loads on the local infrastructure, minimizing the impact of congestion on the larger community.

EDUCATION

Provide students with buses and encourage car-pooling to reduce the load on local traffic. Bus stops and carpool pick-up/drop-off areas should be secure and located where they are easily monitored by teachers and administrators.

MOSQUES

Ensure that existing roads and intersections have the capacity to handle future traffic loads from the mosque, especially during peak times. Peak times include hours of prayer services, Friday midday, and evenings throughout the holy month of Ramadan.

SPORTS

Provide adequate points of exit from the facility to reduce vehicle exit time for peak events. Strategies such as contra flow lanes and stoplight manipulation can be used to direct traffic out of the facility's parking areas. Reduction of exit times can reduce vehicle idle time, carbon emissions, and user comfort. Additionally, exit times may be reduced by timing large scale events to take place during regular non-peak traffic times.

FURTHER RESOURCES

Websites:

1. United States. "Urban Sustainability & the Built Environment." *Sustainability*. US Environmental Protection Agency. Web. 11 June 2010. <http://www.epa.gov/sustainability/builtenvironment.htm>.

2. *Institute of Transportation Engineers*. Institute of Transportation Engineers, 2010. Web. 06 August 2010. <http://www.ite.org/>.

3. Kallman, Matt. "August 2008 Monthly Update: Air Pollution's Causes, Consequences and Solutions" *EarthTrends*. World Resources Institute, 20 August 2008. Web. 12 July 2010. <http://earthtrends.wri.org/updates/node/325>.

4. Federal Highway Administration "Traffic Congestion and Reliability: Trends and Advanced Strategies for Congestion Mitigation" U.S. Deptartment of Transportation, 10 November 2010. Web 10 March 2011. < http://ops.fhwa.dot.gov/congestion_report/chapter2.htm>

Publications:

1. United Arab Emirates. Abu Dhabi Urban Planning Council. *Abu Dhabi Urban Street Design Manual*. Abu Dhabi: UPC, 2009. Print.

[UC.3] Public Transportation

SCOPE

COMMERCIAL | CORE + SHELL | RESIDENTIAL | EDUCATION | MOSQUES | HOTELS | LIGHT INDUSTRY | SPORTS

- Rated for SINGLE and GROUP RESIDENTIAL

DESCRIPTION

Encourage development near access to effective public transportation networks in order to reduce private transportation needs.

GUIDELINES

Select a site that is in close proximity to existing transportation networks to encourage the use of public transportation and reduce the need for private transportation. Transit stops should be easily accessible from the project site by pedestrians and cyclists. Where possible, provide direct paths and walkways from main entrances to nearby transit stops. Paths and walkways should be clearly marked and shaded from direct sunlight to encourage the use of public transportation and ensure convenient access for pedestrians and cyclists.

Transportation services should be assessed for adequate frequency, especially during peak hours of the day. Study the projected usage patterns of future building users, and ensure that the schedule and frequency of the transportation services will accommodate future needs.

Plan to directly connect varying routes of travel to ensure access to transportation leaving the project site. If a public transport stop is not located within walking distance of the building entrance, provide shuttle services from the project site to a nearby transport stop. The schedule and frequency of the shuttle service should accommodate the projected needs of the building users.

Consult with local planning authorities to determine alternative sustainable transport solutions for users of the project. Future transportation networks that are planned and funded by the completion of the project may also be considered for this criterion. Encourage roadway designers to designate sections of road for public transportation in order to expedite travel, promote the functionality of public transportation, and encourage the use of alternative fuels. Consider alternate types of transportation such as buses, street car trolleys, subways, and trains/railways, to provide efficient and expedited travel between longer distances. If possible, locate stations within or near the project site or major transit hubs. If possible, provide informational guides or shuttle transportation to and from the project site to the nearest transit station. Separating railway transportation either above or below roadways, through elevated or subterranean structures, can alleviate traffic congestion and expedite both private and public transit. Create rate programs that reduce fees depending upon the duration of the pass or the number of trips per pass. Consider creating incentive programs to encourage the use of public transportation, such as reduced satellite parking fees and subsidized or complimentary transit passes, for regular building employees.

HOTELS

Ensure nearby public transportation can support the capacity of the hotel staff and guests. If possible, provide shuttle transportation from the hotel to nearby transit hubs and airports.

SPORTS

Transportation services should be assessed for adequate frequency, especially during peak events and peak commuter hours. Ensure nearby public transportation services can support the capacity of the sport facility staff and spectators, especially during peak event times.

FURTHER RESOURCES

Websites:

1. United States. Environmental Protection Agency. *Office of Transportation and Air Quality of the US EPA*. Web. 10 August 2010. <http://www.epa.gov/otaq/>.

Publications:

1. British Council for Offices. *A Good Practice Guide to Green Travel Plans*. London: BCO, 2004. Print.

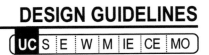
[UC.4] Private Transportation

SCOPE

COMMERCIAL | EDUCATION | HOTELS | LIGHT INDUSTRY | SPORTS

DESCRIPTION

Reduce the use of private transportation to ease the impact on traffic congestion and harmful emissions.

GUIDELINES

The parking capacity on the site should meet, but not exceed, minimum zoning requirements. Determine the expected number of cars on the site based on the maximum number of building users at any time. Parking needs can be reduced by providing means and incentives for alternative transportation including shuttle services, preferred parking for car- or van-pooling, parking fees, transit passes, convenient bicycle racks, changing/showering facilities for bicyclists, and bike sharing programs. Encourage the use of low-emission vehicles, including hybrid vehicles or vehicles that use electricity or compressed natural gas (CNG). Provide priority parking and charge points on the project site if electric vehicles are used.

Reducing the amount of parking on-site will decrease car usage and encourage building users to plan for alternative means of transportation. However, if alternate provisions for transportation are not available, surrounding neighborhoods or districts may be burdened by on-street parking. The amount of parking spaces available to building users can have a direct impact on adjacent sites. Therefore, the project design should ensure an appropriate allocation of parking spaces. If possible, projects should share parking with adjacent properties to minimize the amount of parking on-site—especially in the case of cluster developments.

If the project provides alternative transportation vehicles such as shuttles or delivery vehicles, the transportation fleet should use alternative fuel sources such as biodiesel, electric, ethanol, or hydrogen. Renewable fuels will reduce emissions and decrease demand on nonrenewable fuel sources. Provide preferred parking for vehicles using alternative fuels and encourage building occupants to use shuttle services.

Encourage the use of bicycles during the cooler months of the year by providing secure bicycle parking near building entrances. Consider providing changing rooms and showering facilities within the new building to promote bicycling and pedestrian activity among building occupants. Design the plan to provide bicycle paths and pedestrian walkways as links between the new project and buildings on adjacent sites, especially those that contain amenities and services. Ensure that paths are direct, safe, and adequately lit without disturbing the nighttime environment.

FURTHER RESOURCES

Websites:

1. United States. Environmental Protection Agency. *Office of Transportation and Air Quality of the US EPA.* Web. 10 August 2010. <http://www.epa.gov/otaq/>.

2. American Council for an Energy-Efficient Economy. Web. 10 August 2010. <http://www.greenercars.com>.

3. Union of Concerned Scientists. *Union of Concerned Scientists Clean Vehicle Program.* Web. 10 August 2010. <http://www.ucsusa.org/clean_vehicles>.

Publications:

1. British Council for Offices. *A Good Practice Guide to Green Travel Plans.* London: BCO, 2004. Print.

2. United Kingdom. Department for Transport. *The Essential Guide to Travel Planning.* London: Department for Transport (DfT), 2008. Print.

3. Union of Concerned Scientists. *Union of Concerned Scientists Clean Vehicle Program.* Web. 10 August 2010. <http://www.ucsusa.org/clean_vehicles>.

[UC.5] Sewer & Waterway Contamination

SCOPE

COMMERCIAL | CORE + SHELL | EDUCATION | MOSQUES | HOTELS | LIGHT INDUSTRY | SPORTS

DESCRIPTION

Avoid contamination of waterways to reduce the burden on public treatment facilities.

GUIDELINES

Develop a Sewer & Waterway Contamination Plan to collect and remove toxic or harmful materials, including solids, sludge, sediment, floating debris, oil, detergents, pool and spa chemicals, pesticides, herbicides, and scum from stormwater or wastewater discharges, to prevent contaminants from reaching public utilities. Management practices should include pollution prevention as well as treatment devices and methods. The degree and type of treatment will vary dependent on the specific soil conditions, the building use, and the amount of precipitation on the site. Consider utilizing source control systems and oil separators where necessary to prevent sewer and waterway contamination.

Runoff drains located in areas that have a low risk of surface water pollution should specify surface water management measures, such as permeable surfaces, filter drains, sand filters, swales, filter strips, and infiltration devices, to prevent contamination of waterways. These methods treat surface water using natural processes of physical filtration, sedimentation, adsorption into materials and soils, and biological degradation. The degree of treatment varies with each system and should be selected based on the specific needs of the project, including the amount of water available on the site for such systems. The various forms of biofiltration are likely to require larger quantities of water to function effectively, and so it is recommended to select other forms of filtration devices to decontaminate runoff. Collect, store, and reuse runoff where possible to conserve water on the site and reduce the burden on public treatment facilities.

Runoff drains located in areas that have a high risk of surface water pollution from substances, such as oil and petrol, should specify oil/petrol separators or their equivalent to minimize the risk of further contamination. Paved areas should be cleaned regularly to reduce pollution from oil, gasoline, and other automotive fluids. Exterior areas in the project that contain waste or recycling facilities should be managed and cleaned properly to avoid contaminating waterways with harmful substances.

After the building is complete, test water samples collected from the site to ensure that water is free of contaminants. Testing should be performed quickly and efficiently to avoid the corruption of samples due to changes in equilibrium, temperature, and other biophysical characteristics. Sample containers must be sterilized and made of materials with minimal reactivity with collected materials. Conduct a physical analysis of water samples on-site to ensure accuracy as water samples obtained on-site will exist in equilibrium with the surroundings. Physical analysis of on-site water should measure pH, temperature, total suspended solids (TSS), turbidity, and total dissolved solids (TDS). Additionally, projects must perform a chemical analysis of water samples collected from the site in a qualified laboratory to ensure no toxic substances are present. The chemical analysis will measure levels of dissolved oxygen (DO), nitrate-N, orthophosphates, chemical oxygen demand (COD), biochemical oxygen demand (BOD), pesticides, and metals.

HOTELS

Hotels built on the shoreline or those that have private beaches should ensure that water bodies near the hotel are not contaminated from the large amount of sewage generated by the hotel. Avoid construction directly on the beach in order to avoid disruption of the marine ecosystem and topography.

LIGHT INDUSTRY

The industrial process has the potential to generate toxic or harmful materials that have to be isolated and treated. When a facility handles or generates harmful materials, proper filters and other point-source controls should be utilized as close to the source of pollution as possible. Once these materials are isolated, projects should also have the capabilities to store the materials until they can be properly disposed.

SPORTS

Sports facilities built on the shoreline or those that require ocean or river front competition venues should ensure that water bodies are not contaminated from the large amount of sewage generated during peak and non-peak events. Avoid construction directly on the beach in order to avoid disruption of the marine ecosystem and topography.

FURTHER RESOURCES

Websites:

1. *The Center for Watershed Protection's Stormwater Manager's Resource Center.* Stormwater Manager's Resource Center. Web. 04 June 2010. <http://www.stormwatercenter.net>.

2. Natural Resources Defense Council. Web. 17 August 2010. <http://www.nrdc.org/water/oceans/ttw/faq.asp>.

Publications:

1. Center for Watershed Protection and the Maryland Department of the Environment Water Management Administration. *Maryland Stormwater Design Manual,Volumes I & II*. Baltimore: 2000. Web. 04 June 2010. <http://www.mde.maryland.gov/Programs/WaterPrograms/SedimentandStormwater/stormwater_design/index.asp>.

2. United States. Environmental Protection Agency. *Stormwater Best Management Practice Design Guide*. Washington: EPA, 2004. Print.

3. United States. Environmental Protection Agency. *Managing Storm Water Runoff to Prevent Contamination of Drinking Water*. Washington: EPA, 2001. Print.

4. United Kingdom. Environment Agency / SEPA / Environment & Heritage Service. *General Guide to the Prevention of Pollution: PPG1*. United Kingdom: Environment Agency / SEPA / Environment & Heritage Service. Print.

5. United Kingdom. Environment Agency / SEPA / Environment & Heritage Service. *Use and Design of Oil Separators in Surface Water Drainage Systems: PPG3*. United Kingdom: Environment Agency / SEPA / Environment & Heritage Service, 2006. Print.

6. United Kingdom. Environment Agency / SEPA / Environment & Heritage Service. *Pollution Prevention Pays*. Bristol, UK: Environment Agency, 2004. Print.

7. United Kingdom. Environment Agency. *Groundwater Protection: Policy and Practice (GP3)*. Bristol, UK: Environment Agency, 2007. Print.

8. United Kingdom. *Drain and Sewer Systems Outside Buildings. Hydraulic Design and Environmental Considerations, BS EN 752-4*. London: British Standards Institution, 1998. Print.

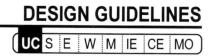
[UC.6] Acoustic Conditions

SCOPE

COMMERCIAL | CORE + SHELL | RESIDENTIAL | EDUCATION | MOSQUES | HOTELS

- Rated for SINGLE and GROUP RESIDENTIAL

DESCRIPTION

Encourage selection of a project site with the least amount of noise exposure.

GUIDELINES

Avoid selecting a site that is affected by high levels of noise pollution generated on nearby sites. Noise pollution occurs in both exterior and interior spaces, and designing quiet spaces is often difficult and may require more material and energy. Projects with facilities that are especially noise-sensitive should not be placed near sites with high levels of noise from equipment or activities such as airport runways or heavy industrial sites. Each project will calculate noise levels for its site taking all surrounding sources and remedies as part of the wider development into account.

There are a number of design elements that can be used to mitigate external noise pollution. Window screens that allow air to pass through can provide some reduction in noise pollution. A portico is also an effective design element that can reduce the amount of noise that enters the interior spaces. Furnishings and surface treatments, including fabric wall hangings, can be used to further dampen unwanted noise. Additionally, design the interior layout to keep spaces with sensitive acoustic requirements away from roadways and other sources of noise pollution.

RESIDENTIAL | EDUCATION

Residences and educational institutions are especially noise-sensitive and should not be placed near sites with high levels of noise from equipment or activities such as airport runways or heavy industrial sites.

HOTELS

When the location of the hotel necessitates proximity to sites of strong acoustic disturbance, such as airports or highways, consider taking additional measures to mute the noise. Airport hotels are especially prone to noise pollution and may require additional measures to ensure proper acoustic conditions. The majority of exterior noise is transmitted into guestrooms through the windows and Packaged Terminal Air Conditioner (PTAC) units. It is important to choose exterior windows, doors, and building assembly materials that have high Sound Transmission Class (STC) values because they will reduce the amount of noise intrusion. For best results, it is crucial that building elements designed for acoustic dampening work together because combining acoustically-strong building materials with acoustically-weak windows or doors can compromise the overall effectiveness of the assembly.

FURTHER RESOURCES

Websites:

1. *Site DNL Calculator.* The Environmental Planning Division of the Department of Housing and Urban Development, 2009. Web. 3 June 2010. <http://www.hud.gov/offices/cpd/environment/dnlcalculatortool.cfm>.

Publications:

1. United States. Department of Housing and Urban Development. *Chapter 5, The Noise Guidebook.* Washington: Department of Housing and Urban Development Office of Policy Development and Research, 2006. Web. 3 June 2010. <http://www.hud.gov/offices/cpd/environment/training/guidebooks/noise/index.cfm>.

2. United States. Department of the Navy. *Guidelines for Sound Insulation of Residences Exposed to Aircraft Operations.* Washington: Department of the Navy, Naval Facilities Engineering Command, 2005. Print.

[UC.7] Proximity to Amenities

SCOPE

COMMERCIAL | CORE + SHELL | RESIDENTIAL | EDUCATION | HOTELS | LIGHT INDUSTRY
- Rated for SINGLE and GROUP RESIDENTIAL

DESCRIPTION

Encourage development of near pedestrian accessible amenities in order to reduce transportation needs.

GUIDELINES

Select a site that is in close proximity to many existing local amenities to reduce the need for excess vehicular transportation and to encourage bicycling and pedestrian activity. Building users and occupants should be able to easily and safely access a variety of basic services from the proposed building and/or development.

Conduct a field survey to determine the range of amenities, services, and facilities accessible from the proposed site. Amenities, services, and facilities may include, but are not limited to, the following: bank, convenience store, food outlet/restaurant, place of worship, market, fitness center, community center, commercial offices, medical facilities, administrative offices, cleaners, pharmacy, post office, educational facilities, and child care facilities. The proposed site plan and building layout, including the position of entrances, walkways, and roads, should consider the location of existing amenities and services to ease accessibility.

In cases where a basic range of amenities does not already exist in the nearby vicinity, consider providing these services within the new building and/or development to reduce the need for transportation and encourage the development of the community.

FURTHER RESOURCES

Websites:

1. *Urban Land Institute.* Urban Land Institute, 2010. Web. 09 August 2010. <www.uli.org>.

2. *International Union for the Scientific Study of Population.* IUSSP, n.d. Web. 09 August 2010. <www.iussp.org>.

3. *Urban Sustainability & the Built Environment.* Environmental Protection Agency, 02 February 2010. Web. 11 June 2010. <http://www.epa.gov/sustainability/builtenvironment.htm>.

4. *Smart Growth Online.* Smart Growth Network, 2010. Web. 09 August 2010. <http://www.smartgrowth.org/>.

Publications:

1. Moe, Richard and Carer Wilkie. *Changing Places: Rebuilding Community in the Age of Sprawl.* Henry Holt & Company, 1999. Print.

2. Rocky Mountain Institute and Alex Wilson, et al. *Green Development: Integrating Ecology and Real Estate.* John Wiley & Sons, 1998. Print.

3. United States. Environmental Protection Agency. *Creating Great Neighborhoods: Density in Your Community.* Washington: EPA, 2005. Print.

4. United States. Environmental Protection Agency. *Getting to Smart Growth: 100 Policies for Implementation.* Washington: EPA, 2002. Print.

5. United States. Environmental Protection Agency. *Getting to Smart Growth II: 100 More Policies for Implementation.* Washington: EPA, 2003. Print.

6. United States. Environmental Protection Agency. *Our Built and Natural Environments.* Washington: EPA, 2001. Print.

7. Bernstein, Richard A., Ed. *A Guide to Smart Growth and Cultural Resource Planning.* Madison: Wisconsin Historical Society, Division of Historic Preservation, 2002. Print.

8. Burchell, Robert W. et al. *TCRP REPORT 74: Costs of Sprawl—2000.* Washington: National Academy Press, 2002. Print.

9. Goldberg, David. *Choosing Our Community's Future.* Washington: Smart Growth America, 2006. Print.

10. International City/County Management Association and US Environmental Protection Agency. *This is Smart Growth.* Smart Growth Network, 2006. Print.

[UC.8] Accessibility

SCOPE

COMMERCIAL | CORE + SHELL | EDUCATION | MOSQUES | SPORTS

DESCRIPTION

Encourage site selection near access to existing pedestrian and bicycle pathway networks, as well as intended users, in order to reduce vehicular transportation needs.

GUIDELINES

Projects should enhance mobility by selecting a site that connects to an extensive, user-friendly network of pedestrian and bicycle pathways. Project sites that are well-connected to residences, offices, and amenities promote accessibility and convenience to users. Additionally, increasing pedestrian and bicycle access reduces the demand for vehicular transportation, thereby reducing the harmful emissions that adversely affect human health and contribute to global climate change.

Consider the existing pedestrian and bicycle pathways on adjacent sites to ensure appropriate and efficient connectivity between sites. Incorporate sidewalks and bicycle lanes into the design of all new developments or buildings to ensure sufficient pedestrian and bicycle access along circulation routes and from roads around the perimeter of the site. Design the site to include additional pedestrian and bicycle pathways, separate from roadways, to provide direct access for pedestrians and bicyclists between sites of interest. Dedicated pedestrian pathways separate from roadways can create user-friendly open spaces that become points of interest in themselves. Separate pedestrian pathways from roadways with the use of raised sidewalks, curbs, or bollards to clearly identify the paths. Pedestrian pathways can be used in conjunction with the design and layout of plazas, picnic areas, monuments, and other site features.

Pathways should be designed to provide direct and safe connections for pedestrians and bicyclists to maximize convenience to building occupants and other users of the site. If possible, design paths to be visible from other areas on the site to foster a sense of security and promote a safer environment for project users. Additionally, plan pathways to avoid loading zones, mechanical equipment, and other spaces unpleasant for pedestrians. Paths and building entrances should be clearly labeled in order to allow for convenient wayfinding between all entrances and points of interest, as well as to adjacent properties and public transportation nodes. Building frontages and entrances should face the street to promote active pathways and streetscapes. Use design features, including signage, awnings, and prominent entrance areas, to assist in wayfinding and walkability.

Provide pathways with the appropriate surface materials, accessible width and slope, and the appropriate lighting scheme to ensure pedestrian comfort and safety. Surfaces should be solid, smooth and able to withstand inclement weather regardless of the specific material used. Provide sufficient space for the anticipated foot traffic as well as for the presence of other features on the pathways. Pathways should adhere to universal design standards for handicap accessibility such as providing a minimal slope and easily accessible ramps.

Support the bicycle network by providing the proper infrastructure including bicycle parking facilities, showers, and changing spaces. There are four main types of bicycle pathways: a shared lane in a roadway, a dedicated lane in a roadway, a physically separated lane in a roadway, and a pathway independent of any road. While dedicated and separated bicycle pathways are the most effective way to ensure the safety and usability of a bicycle network, these types of paths are not always possible. The project should design a bicycle pathway network that includes different types of bicycle pathways for different road conditions and user demand and allows for maximum access to the site from the surrounding community.

Provide bicycle pathways with the appropriate surfacing to ensure that the pathways are smooth and free of obstacles such as sewer drains, potholes, or other obstructions. Traffic paint should be a bright color and contain reflective pigments to make it more visible at night. Place signage at regular intervals along bicycle pathways in a position where it is visible to both vehicular and bicycle traffic. Particular attention should be paid to accommodating bicycles at intersections, including providing dedicated spaces for bicyclists to wait at traffic lights or well-marked turning lanes. When vehicular traffic has the potential to cross bicycle paths, additional street markings and signage should be used to clearly mark traffic patterns.

When bicycle pathways are not physically separated from vehicular traffic, it is important to use proper signage and lane markings to alert drivers to the presence of bicycles. Traffic paint should clearly mark the bicycle path, and wide paint stripes or a painted buffer between vehicular and bicycle traffic provides a better visible separation. Physical barriers between vehicular traffic and the bicycle pathway are the most effective way to provide a safe corridor for bicycle travel. Examples of barriers that can be used to separate the bicycle lane from vehicular traffic include curbs, bollards, and vegetation.

FURTHER RESOURCES

Publications:

1. United States. U.S. Department of Transportation. Axelson, Peter, et al. *Designing Sidewalks and Trails for Access: Part I of II: Review of Existing Guidelines and Practices.* Washington: U.S. Department of Transportation, Federal Highway Administration, 1999. Web. 29 June 2010. <http://www.fhwa.dot.gov/environment/sidewalks/index.htm>.

2. Access Minneapolis. *Design Guidelines for Streets and Sidewalks.* Minneapolis: City of Minneapolis Public Works, 2008. Web. 29 June 2010. <http://www.ci.minneapolis.mn.us/publicworks/trans-plan/DesignGuidelines.asp>.

3. Task Force on Geometric Design. *Guide for the Development of Bicycle Facilities.* Washington: American Association of State Highway and Transportation Officials, 1999. Web. 29 June 2010. <http://www.sccrtc.org/bikes/AASHTO_1999_BikeBook.pdf>.

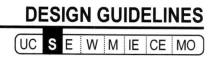
SITE [S]

The design of the proposed project has a direct impact on both the site of the project as well as any adjacent sites. Sustainable landscaping and site design practices can improve the quality of the existing site and landscape, as well as minimize impacts on the surrounding environment including climate change, fossil fuel depletion, water depletion and pollution, air pollution, land use and contamination, and human comfort and health. Consider the following factors when designing the project: the preservation of land, water bodies, and habitats on the site; developing a landscaping plan that encourages the use of native vegetation, reverses desertification, and prevents rainwater runoff; reducing the increase in the heat island effect; reducing the negative effect on pedestrian level air flows; minimizing the impact on the surroundings due to noise pollution, light pollution, and blocked solar potential to adjacent properties; reducing the parking footprint; developing shaded, safe, and effectively illuminated public spaces and pathways; and designing for mixed uses to reduce travel.

Landscaping plans in a hot, dry climate should strive to reduce heat, glare, and desiccation to conserve moisture, and to control temperature for human comfort and horticultural success. In such climates, it is important to design for water-efficient landscaping, to be mindful of conserving water, and to protect the environment. The term "Xeriscape" [coined and copyrighted by Denver Water Department, 1981] encapsulates this idea through seven principles/guidelines:

1. Planning and Design:

The microclimate, topography, vegetation, and soil conditions of the existing site should be considered when developing a landscaping plan. Create an efficient landscaping plan by grouping vegetation by soil, solar, and water needs.

2. Soil Analysis, Preparation, and Improvement:

Soils should be tested to determine the organic matter content and nutrient and acidity levels. Soil quality can be improved in many ways such as by adding organic matter to increase fertility.

3. Practical Turf Areas

Limit the amount of lawn/turfgrass used on the site, as they require a large quantity of water to survive. Where turfgrasses are used, specify species that require less water and are adapted to the region. Long, narrow areas of lawn are difficult to mow and irrigate efficiently and should be avoided.

4. Appropriate Plant Selection

Preserve as much existing healthy vegetation on the site as possible. Native and/or adapted trees, plants, and groundcover should be specified for the landscape plan as they may require little water, pest control, and/or fertilizers.

5. Efficient Irrigation

Inefficient irrigation techniques can waste water through runoff and evaporation. Take into account the soil conditions, plant groupings, microclimate, and topography when selecting the appropriate irrigation technique. Avoid using sprinkler irrigation and other above-ground systems where possible; instead consider utilizing drip irrigation systems which slowly apply water to the plant's roots and have little chance of waste.

6. Use of Mulches/Gravel/Stone

Prevent evaporation and moisture-loss from the soil around plants through the use of materials such as mulch, gravel, or stone. Using these materials in plant beds can also reduce weed growth, help maintain moderate soil temperatures, and prevent erosion from runoff.

7. Appropriate Maintenance

Appropriate, timely maintenance will promote the healthy longevity of a vegetated landscape in a hot, dry climate. Limit the use of harmful fertilizers, regularly remove weeds, and prune bushes and shrubs as needed. Conserve water by maintaining the proper height of grass and checking for irrigation system efficiency.

These principles are intended as a guide to designing and maintaining a landscape plan that conserves water and in turn, protects the environment.

Criteria in this category include:

S.1	Land Preservation
S.2	Water Body Preservation
S.3	Habitat Preservation
S.4	Vegetation
S.5	Desertification
S.6	Rainwater Runoff
S.7	Heat Island Effect
S.8	Adverse Wind Conditions
S.9	Noise Pollution
S.10	Light Pollution
S.11	Shading of Adjacent Properties
S.12	Parking Footprint
S.13	Shading
S.14	Illumination
S.15	Pathways
S.16	Mixed Use

FURTHER RESOURCES

Websites:

1. *Xeriscape Colorado.* Colorado WaterWise, 2005. Web. 04 June 2010. <http://coloradowaterwise. org/XeriscapeColorado/>.

2. *Irrigation Association.* Irrigation Association, 2010. Web. 04 June 2010. <http://www.irrigation.org>.

3. *Water-Efficient Gardening and Landscaping.* University of Missouri Extension, 1993. Web. 04 June 2010. <http://muextension.missouri.edu/xplor/agguides/hort/g06912.htm>.

Publications:

1. United States. Environmental Protection Agency. *Water-Efficient Landscaping: Preventing Pollution & Using Resources Wisely.* Washington: EPA, 2002. Print.

[S.1] Land Preservation

SCOPE

COMMERCIAL | CORE + SHELL | RESIDENTIAL | EDUCATION | MOSQUES | HOTELS | LIGHT INDUSTRY | SPORTS

• Inherited for SINGLE and GROUP RESIDENTIAL

DESCRIPTION

Encourage development on land that is contaminated, previously developed, or has low ecological value. In addition, preserve or enhance the site through remediation, conservation, and/or restoration.

GUIDELINES

Projects should select sites that have been previously developed in order to preserve undeveloped, natural land. The conservation of the natural state of the site should take into consideration the existing topography, soil, trees, plants, groundcover, water features, and wildlife habitats.

Large-scale developments also present an opportunity to remediate and build on land that has been contaminated through industrial waste or other human activities, in order to protect against further contamination of the surrounding region. Furthermore, it is important to maintain, restore, or improve the land in order to combat desertification, prevent the further decline of water quality, and preserve overall ecological health.

Preserving, restoring, or enhancing a site's soil conditions will help to ensure healthy vegetation and wildlife habitats. The Site Assessment Report should identify critical areas in the development such as steep slopes, areas of high water flow, and vegetative or wetland buffers. Additionally, the report should identify the depth and quality of existing soils and determine appropriate remediation or conservation techniques. Using the recommendations of the Site Assessment Report, create a soil management plan detailing strategies to maintain, restore, or improve the land. Strategies may include protecting zones where existing soil and vegetation will not be disturbed, identifying zones of lower quality that will be enhanced with organic material, and preventing soil erosion and compaction.

Integrate organic waste management with landscape management during the operations phase of the building to ensure continued sustainable soil practices. When necessary, import higher quality topsoil to mix with existing soil or to replace soil of lower quality. Minimize soil compaction by identifying pathways and areas during construction for heavier equipment, in order to localize affected areas. Minimize the length of time soil remains barren or uncovered to avoid erosion due to wind or water. Use groundcover, mulch, and/or sand berms in landscaped areas to prevent soil movement.

Use of Previously Developed Land

Selecting a site that has been previously developed is highly encouraged. Building on a previously developed site reduces the impact on the environment and prevents more valuable, undeveloped land from being disturbed. In cases where the selected site contains both developed and undeveloped land, the footprint of the proposed development should occupy those areas that have been previously developed.

Previously developed land includes buildings, roads, parking lots, or land that has been graded or altered directly by human activities. All fixed structures and surfaces to be constructed within the site boundaries are considered part of the proposed development including buildings, hardscapes, parking lots, roads, and pathways. Temporary structures and surfaces, such as construction-related offices, storage, parking, and roads, can disturb sensitive land areas and, if possible, should be sited on previously developed land.

Use of Contaminated Land

A contaminated land specialist should conduct an investigation to test for hazardous levels of pollution on the site. The specialist will determine strategies to remediate contaminated areas in order to prevent further risks to the environment and to human health. Implementation of the remediation strategies must be completed before construction can begin on the proposed site.

Strategies for remediation should be determined by the type and degree of contamination, natural site features, level of short- and long-term effectiveness, available funds, and time frame for completion. All remediation strategies should have minimal disruption to the site including underground features. The land should continue to be monitored after remediation takes place to ensure that all hazardous substances have been completely cleared from the site.

Contaminated groundwater can be remediated using pump-and-treat technologies, where the water is pumped to the surface and treated using physical or chemical processes. The remediation of contaminated soils can be undertaken through several strategies such as in situ applications, off-site disposal, the use of bioreactors, and solar detoxification technologies. The implications of all remediation strategies should be considered to minimize negative environmental impacts.

Soil Disturbance

Reducing unnecessary disturbance of soil is important for conserving the natural resources of the site. Healthy soils can effectively cycle nutrients, store carbon as organic matter, maximize water holding capacity, and provide a healthy rooting environment and habitat to a wide range of organisms. Excessive soil disturbance can include actions such as excavating for the construction of buildings, landscaping, infrastructure, man-made water bodies, dredging for new coastline, or infilling for man-made islands. Excavation or fill required on-site will not only decrease the ecological value of the site, but will also increase the need for transport and contribute to the depletion of fossil fuels. Soil disturbance may also release natural Volatile Organic Compounds (VOCs), which leads to air pollution. Additionally, the excavation of soil to create man-made water bodies requires a significant amount of water and therefore

increases the demand for sea water desalination.

Projects should minimize the amount of soil that is transported into or out of the site, as well as design sites that take advantage of natural features. Limit grading on the site and plan construction machinery routes to minimize the amount of soil compaction.

Erosion Prevention

Preventing erosion is important on the site because it can cause degradation to the habitat of plants and animals and strip the soil of valuable nutrients. Erosion may also cause scour in bearing soils which can undermine and cause instability of both above and below grade structural features, including earthen embankments, built structures, and roadways. Projects should develop a Soil Erosion Plan to employ erosion control practices such as preserving natural vegetation where possible, directing the runoff away from exposed soils, planting temporary groundcover, and permanent revegetation of areas at risk for erosion damage.

Projects should avoid soil disturbance as the best method for minimizing erosion. Erosion rates are directly proportional to the type and density of groundcover on the site. Preservation of the natural vegetation is the most efficient and inexpensive form of erosion control, greatly reducing the need for revegetation. Disturbed areas require additional means of erosion prevention and sediment control, because they are more prone to erosion and invasive weed species.

Projects should create buffer zones and setbacks to reduce the amount of erosion and runoff from the site. Restrict activities in areas with erosive potential by creating undisturbed areas with natural vegetation or areas that are suitable for revegetation with native plant species. Buffer zones and setbacks may be used to protect streams and waterways, environmentally sensitive habitats, neighboring properties, structures, roadways, and pathways. Buffer zones and setbacks are low-maintenance and easy to visually inspect. Additionally, they aid with the filtration of sediment and absorption of runoff, as well as provide habitats for native flora and fauna.

Non-vegetated areas prone to erosion may be replanted with native species to prevent further damage. Revegetation will help prevent erosion by slowing down runoff drainage on hillsides and protecting soil from wind erosion. The roots of plants serve to stabilize soils, and revegetation enhances water infiltration in the soil, reduces runoff, and traps sediment.

LIGHT INDUSTRY

Site Impact Over Time

Industrial processes have the potential to impact the quality of the land over time. The use of hazardous materials, the presence of industrial equipment, and the general industrial processes that take place on-site increase the likelihood that the site can become more contaminated over time. Projects should identify the potential sources of contamination and ways to remediate the hazards over the lifetime of the project.

Continued maintenance of containment and treatment systems is necessary to mitigate potential negative environmental impacts that may occur. Projects should also develop plans for handling accidents and other emergency situations that can expose people and the site to environmental hazards.

FURTHER RESOURCES

Websites:

1. *Institute of Ecology and Environmental Management.* Institute of Ecology and Environmental Management. Web. 04 June 2010. <http://www.ieem.net/>.

2. *Royal Botanic Gardens Melbourne.* Australian Research Centre for Urban Ecology, 21 May 2010. Web. 04 June 2010. <http://arcue.botany.unimelb.edu.au/>.

3. National Center for Environmental Research. *Implementation of Green Roof Sustainability in Arid Conditions.* Environmental Protection Agency, 28 October 2008. Web. 04 Jun 2010. <http://cfpub. epa.gov/ncer_abstracts/index.cfm/fuseaction/display.abstractDetail/abstract/8850/report/0>.

4. Lockheed Martin Environmental Services. *Effects Of Soil Disturbance On VOC Emissions From Aged, Contaminated Soil.* Environmental Protection Agency, 4 September 2003. Web. 29 Jul 2010. <http://www.epa.gov/esd/cmb/qa/bs103_QA.pdf>.

Publications:

1. S. Coventry, M. Kingsley, and C. Woolveridge. *Environmental Good Practice - Working on Site, C503.* United Kingdom: Construction Industry Research and Information Association, 1999. Print.

2. Institute of Ecology and Environmental Management. *Guidelines for Ecological Impact Assessment in the United Kingdom.* IEEM, 26 June 2006. Web. 04 June 2010. <http://www.ieem.net/ecia/>.

3. State of Qatar. *National Biodiversity Strategy and Action Plan, State of Qatar.* Doha: October 2004. Web. 04 June 2010. <http://www.cbd.int/doc/world/qa/qa-nbsap-01-en.doc>.

4. British Standards Institution. *Code of Practice for Site Investigations, BS 5930.* United Kingdom: British Standards Institution, 1999. Print.

5. British Standards Institution. *Soil Quality - Sampling, Part 1: Guidance on the Design of Sampling Programmes, BS ISO 10381-1.* United Kingdom: British Standards Institution, 2002. Print.

6. California Department of Transportation. *Key Concepts Of Sustainable Erosion Control: Technical Guide.* Sacramento: California Department of Transportation, 2010. Print.

7. Tehachapi Resource Conservation District et al. *Erosion Control Guide for Desert and Mountain Areas.* Rosemead: Southern California Edison. 2010. Print.

8. United States. Federal Emergency Management Association. *Erosion, Scour, and Foundation Design.* Washington: Federal Emergency Management Association. 2009. Print.

[S.2] Water Body Preservation

SCOPE

COMMERCIAL | CORE + SHELL | RESIDENTIAL | EDUCATION | MOSQUES | HOTELS | LIGHT INDUSTRY | SPORTS

• Inherited for SINGLE and GROUP RESIDENTIAL

DESCRIPTION

Encourage development that prevents or minimizes ecological degradation to water bodies in order to preserve the natural resources of the region.

GUIDELINES

Projects should protect all natural water bodies on the site, including coastlines and groundwater, to prevent degradation to these limited resources. Coastline development not only affects the quality of nearby water bodies, but also the overall ecological health of the habitats dependent on them. Furthermore, in a region with limited precipitation and shrinking groundwater levels, it is especially important to conserve the remaining amount of naturally available fresh water. To maintain the direction of groundwater flow, protect existing trees and significant vegetation, and retain natural slopes and the topography of the site.

Water bodies are defined as any area that holds surface or groundwater including, but not limited to, streams, rivers, lakes, estuaries, bays, gulfs, or aquifers. These water bodies are home to a variety of flora and fauna.

New developments should reduce activities that have the potential to harm the ecological diversity of water bodies. Avoid infilling, which requires dredging areas in or near bodies of water to create new land. Dredging creates excess silt and debris in existing water bodies in addition to destroying marine life and habitats. Avoid dredging on the site in order to protect coastlines and soils in the gulf. Also, precautions should be taken to ensure that pollutants from runoff or direct dumping from construction and industrial projects do not contaminate the water supply.

Preserve all water bodies by developing a Water Body Preservation Plan to collect and remove all toxic or harmful materials in order to prevent contaminants from reaching waterways. Specify surface water management measures such as permeable surfaces, filter drains, sand filters, swales, filter strips, and infiltration devices for runoff drains located in areas that have a low risk of surface water pollution. These methods treat surface water using natural processes of physical filtration, sedimentation, biological degradation, and absorption into materials and soils. The degree of treatment varies with each system and should be selected based on the specific needs of the project, including the amount of water available on-site for these systems. Methods of biofiltration are likely to require larger quantities of water to function effectively; therefore, it is recommended to use other filtration devices to decontaminate runoff. Collect, store, and reuse runoff where possible to conserve water on the site and reduce the burden on public treatment facilities.

Water bodies are also preserved through adherence to coastal protection regulations such as a mandated buffer between the boundaries of a project site and the water body. These regulations protect the coast from erosion and serve as a natural filter to limit the contaminants that reach the water. Adhere to these regulations for all existing coastal areas and protected habitats.

HOTELS

Hotels with private beaches should ensure that these areas are not destructive to habitats, and shoreline impact should be minimal. Avoid construction near the shoreline to avoid disturbing natural ecological barriers and prevent water body contamination.

FURTHER RESOURCES

Websites:

1. Venturini, Kate. *Coastal Buffer Zones.* Ecological Landscaping Association, 15 Jan 2010. Web. 24 Jun 2010. <http://www.ecolandscaping.org/news/?p=59>.

2. *Water Resources and Freshwater Ecosystems – Qatar.* World Resources Institute, 2003. Web. 24 June 2010. <http://earthtrends.wri.org/pdf_library/country_profiles/wat_cou_634.pdf>.

3. *Coastal and Marine Ecosystems – Qatar.* World Resources Institute, 2003. Web. 24 June 2010. <http://earthtrends.wri.org/pdf_library/country_profiles/coa_cou_634.pdf>.

4. *Economic Sustainability: Indicators of Coastal Water Quality.* Columbia University Socioeconomic Data and Applications Center, 1997-2010. Web. 24 June 2010. <http://sedac.ciesin.columbia.edu/es/seawifs.html>.

Publications:

1. Sheppard, Charles, Andrew Price, Callum Roberts. *Marine Ecology of the Arabian Region, Patterns and Processes in Extreme Tropical Environments.* London: Academic Press, 1992. Print.

2. United Nations. Economic and Social Commission for Western Asia. *ESCWA Water Development Report 2, State of Water Resources in the ESCWA Region.* New York: United Nations, 2007. Print.

3. Alsharahan, A.S., et al. *Hydrogeology of an Arid Region: the Arabian Gulf and Adjoining Areas.* Amsterdam: Elsevier, 2001. Print.

[S.3] Habitat Preservation

SCOPE

COMMERCIAL | CORE + SHELL | RESIDENTIAL | EDUCATION | MOSQUES | HOTELS | LIGHT INDUSTRY | SPORTS

• Inherited for SINGLE and GROUP RESIDENTIAL

DESCRIPTION

Encourage development that preserves and/or enhances the biodiversity of the site in order to protect the natural ecosystems of the region.

GUIDELINES

Projects should protect all habitats, natural vegetation, and wildlife on the site to prevent degradation to these limited resources. Alterations to the terrain can negatively impact the region's ecosystems and biological diversity. Ecologically sensitive habitats are not limited to the project site and usually extend into adjacent areas and beyond. Therefore, actions taken within the project site should consider consequences that extend well past its own boundaries.

Habitat preservation maintains the ecological balance in a region and helps protect and increase the level of biodiversity. Areas of the site that have been identified as ecologically relevant or valuable should be preserved per the recommendations from the Site Assessment Report. Examples of sites with high ecological value are: areas containing rare and endangered wildlife species; sites with a high representation of indigenous biodiversity; key biological sites such as wetlands, seagrass beds, and mangroves; and areas that could be easily rehabilitated to provide a suitable habitat for wildlife. The project should not decrease the habitat for endangered or threatened species of flora and fauna. Where applicable, wildlife corridors should also be protected.

Designate a buffer zone to protect habitats from the design and construction process. In cases where habitats and vegetation are to be disturbed during construction, develop a plan to restore the native ecology by replanting the disturbed vegetation and reintroducing the same species and habitats after construction is complete. During the design phase projects should reduce the project footprint by sharing facilities, access roads, and walkways within clustered developments and with existing and future buildings on adjacent sites.

Consult with an ecologist to create a Habitat Preservation Plan that maintains and enhances habitats and ecosystems on the site. The plan should catalog all species on-site before and after construction in order to preserve biodiversity and encourage the use of native plants. Consult [S.4] Vegetation for a more detailed explanation of appropriate planting and landscaping design.

FURTHER RESOURCES

Websites:

1. *Institute of Ecology and Environmental Management.* Institute of Ecology and Environmental Management. Web. 04 June 2010. <http://www.ieem.net/>.

2. *Royal Botanic Gardens Melbourne.* Australian Research Centre for Urban Ecology, 21 May 2010. Web. 04 June 2010. <http://arcue.botany.unimelb.edu.au/>.

3. *Ecosystem Management.* United Nations Environment Programme. Web. 30 June 2010. <http://www.unep.org/ecosystemmanagement/>.

Publications:

1. S. Coventry, M. Kingsley, and C. Woolveridge, *Environmental Good Practice - Working on Site, C503.* United Kingdom: Construction Industry Research and Information Association, 1999. Print.

2. Institute of Ecology and Environmental Management. *Guidelines for Ecological Impact Assessment in the United Kingdom.* IEEM, 26 June 2006. Web. 04 June 2010. <http://www.ieem.net/ecia/>.

3. State of Qatar. *National Biodiversity Strategy and Action Plan, State of Qatar.* Doha: October 2004. Web. 04 June 2010. <http://www.cbd.int/doc/world/qa/qa-nbsap-01-en.doc>

4. British Standards Institution. *Code of Practice for Site Investigations, BS 5930.* United Kingdom: British Standards Institution, 1999. Print.

5. British Standards Institution. *Soil Quality - Sampling, Part 1: Guidance on the Design of Sampling Programmes.* BS ISO 10381-1. United Kingdom: British Standards Institution, 2002. Print.

[S.4] Vegetation

SCOPE

COMMERCIAL | CORE + SHELL | RESIDENTIAL | EDUCATION | MOSQUES | HOTELS | LIGHT INDUSTRY | SPORTS

- Inherited for SINGLE RESIDENTIAL
- Rated for GROUP RESIDENTIAL

DESCRIPTION

Minimize lawn and encourage native or low-impact vegetation for the site in order to reduce irrigation demand.

GUIDELINES

Any areas of the proposed site that are not occupied by buildings and roads should be vegetated and shaded to the greatest extent possible. Projects should select native vegetation and reduce lawn area to lower irrigation demand and increase the long-term viability of the landscaping. Additionally, native vegetation that is non-invasive and adapted to the local climate supports biodiversity, combats desertification, and helps to reduce the carbon footprint of the project.

Preserve and reuse all existing healthy trees and plants within the site boundary, and develop a landscaping plan that provides for an appropriate amount of new vegetation on the site. Native and adapted plants are indigenous to a specific region or have been adapted to the local climate. They have a higher chance of survival than exotic species and provide habitats for local wildlife. Furthermore, native plant species require nominal maintenance, minimal irrigation, and little to no chemical inputs such as fertilizers, pesticides, and herbicides. Local and regional governmental agencies and/or consultants should be able to provide a list of approved and recommended tree and plant species that are appropriate for the region.

Refer to the tables on the following pages for a list of recommended trees, shrubs, vines, and groundcovers well-suited for hot sub-tropical climates, though not necessarily native to Qatar. The examples included below are by no means inclusive of all possible recommendations, and if necessary, each project should conduct their own research for additional options. Plant species listed as 'dry condition' and 'minimum' are the hardiest plants for a desert environment and require the least amount of water. Plants that are labeled as having 'moderate' water requirements may necessitate some irrigation, but their inclusion is meant to provide designers with a wider selection of species that are appropriate for the region. Weigh the benefits and potential drawbacks of selecting plants that may require more maintenance and irrigation to thrive in a desert environment. For a list of plant species native to Qatar, refer to the report entitled "Qatar: Country Report to the FAO International Technical Conference on Plant Genetic Resources".

In a region where fresh water contains a large amount of embodied energy, it is important to consider the amount of water the landscape design demands. Lawns, unlike native plant landscapes, tend to require large quantities of water, labor, and sometimes chemicals. Due to their high ecological cost, lawns should be minimized or avoided. Instead of planting grass, consider plant species that have minimal water demand and require less maintenance. Develop a landscaping strategy to minimize the amount of lawn used as vegetative cover and increase the amount of vegetation that is ecologically sensitive to an arid climate.

Landscaped areas should provide adequate soil depth, proper ventilation, and appropriate levels of sunlight to encourage healthy vegetation. Invasive plants, noxious weeds, and any other vegetation that could be destructive to the site should be avoided. Use gravel, mulch, and/or stones around plants to protect the soil from solar heat gain and water loss.

Utilize efficient irrigation techniques to provide vegetation with the appropriate amount of water. While spray irrigation is an efficient and easy method of watering, it can be wasteful in hot climates as some of the water will evaporate before reaching the ground. Drip irrigation systems create no overspray, blocked spray, or runoff and can be installed on top of soil or below the surface to minimize water loss due to evaporation. Drip irrigation systems are very efficient in terms of water conservation because water is introduced directly to the roots of the plant. Irrigation systems can also be designed to collect and take away excess water from the plants below grade in order to recycle and reuse as much water as possible.

Additionally, the project may consider the use of vegetated roofs to enhance the site and building. Besides providing green spaces to occupants above ground level, vegetated roofs can provide habitats for wildlife and create additional shade for the project and building users. In addition, the benefits of vegetated roofs include reduced energy costs, reduced heat island effect, extended roof life, insulation, and sound absorption.

Refer to the principles of xeriscape landscaping found at the beginning of this section for further recommendations.

RESIDENTIAL

Encourage publicly accessible green spaces within residential developments by providing vegetated and shaded areas to promote community development and recreation. These spaces allow for outdoor activity, community interactions, and educational environments improving the health and well-being of the community. Additionally, these spaces provide ecological benefits such as reduced heat island effects, greater stormwater control, and wildlife habitats. Locate residential developments in close proximity to existing publicly accessible spaces to reduce the need to travel for outdoor recreation.

HOTELS | SPORTS

Recreation areas and competition courses that are heavily landscaped, such as golf courses and playing fields, should take particular care in managing lawns and landscaped areas. For lawn areas, use saline tolerant grass species that require minimal water and can make use of brackish water for irrigation. Also, projects should try to reduce lawn areas to only target fairways, putting greens, or areas necessary for competition, and alternate grass species according to seasonal tolerance. The need for irrigation will be reduced by designing smaller areas of lawn and installing water retention systems underneath the soil.

During the evening, fairways, putting greens, and playing fields may be covered with clear tarps to catch and redistribute condensation to further reduce irrigation needs. Driving ranges should use lawn only on tee areas, or combine lawn tees with synthetic mat tees to eliminate excess lawn. Design driving range practice fields to use the existing, non-manicured desert landscape or a combination of rocks, sand, and low-maintenance native vegetation. For other areas that do not require specific playing surface conditions, use native vegetation that requires minimal maintenance and water.

Trees	Height	Growth Rate	Water Requirements	Salinity Tolerance	Uses	Cultivation
ACACIA [Acacia arabica]	9m	moderate	minimum	8,000 ppm	Grows well near the sea and provides light shade. Can be made to form an impenetrable barrier.	Tolerates very dry conditions. Needs careful pruning to attain a nice shape.
SIRIS [Albizia lebbek]	15-18m	fast	moderate	6,000 ppm	Gives excellent shade. Ideal as a street tree, and planting in large gardens and urban parks.	Shallow-rooted and blows over in windy sites.
WHISPERING PINE [Casuarina equisetifolia]	24.5m	fast	minimum	20,000 ppm	Excellent for planting near the sea, especially in sandy soils. Can be used as a hedge. Provides good shade at maturity.	Tends to grow obliquely in windy locations. Needs good staking and full sun. Heavy feeder and understory will not grow easily.
RED GUM [Eucalyptus camaldulensis]	18m	fast	minimum	9,000 ppm	Good for wide streets, wide avenues, parks, and shelterbelts. Provides good shade at maturity.	Drought resistant. Should be protected from sea winds. Tolerates salinity.
BANYAN [Ficus bengalensis]	18m	moderate	moderate	4,500 ppm	Enormous shade coverage. Ideal for parks and large grounds.	Has a very long life. Grows well in full-sun, moist conditions, and high atmospheric humidity.
JACARANDA [Jacaranda acutifolia]	12m		High drought tolerance. Once established, needs only occasional water	not salt tolerant	Reaches very sizable proportions, and is unsuitable for small properties. Good for urban, street, and boulevard planting.	Full Sun. Drought-tolerant and suitable for xeriscaping. Prefers enriched, sandy, and well drained soils but is tolerant of most soil types.
HORSERADISH TREE [Moringa oleifra]	6-7.5m	moderate	moderate		Small tree and good for light shade.	Well-sheltered and full-sun locations.
OLIVE [Olea europaea]	8-15m	very slow	moderate water, drought tolerant	tolerates salty air	Flourishes on limestone slopes and coastal climate conditions.	Prefers calcareous soils however poor or dry. Tolerates drought well. Will grow in any light soil and even on clay, if well drained.
DATE PALM [Phoenix dactylifera]	18m	slow	dry condition	20,000 ppm	Thrives near the sea. Highly decorative and ideal for avenues or streets.	Tolerates dry conditions but prefers to have its root system within reach of the water table.

Table S.4.1   Plant Species Suitable for Sub-tropical Climates

Trees (con't)	Height	Growth Rate	Water Requirements	Salinity Tolerance	Uses	Cultivation
MANILA TAMARIND [Pithecolobium dulce]	12m	moderate	moderate		Suitable for parks, large gardens, and streets. Effective shade at maturity. Can be used as a hedge or screen and made into impenetrable barrier.	Will grow near the sea. Needs staking and careful pruning to develop a nice shape.
COUNTRY ALMOND [Terminalia catappa]	18m	moderate	minimum	4,500 ppm	One of the best trees for shade. Ideal for streets, urban parks, and courtyards. Can be planted close to buildings where heavy shade is desired.	Grows in a range of soil conditions, even close to the sea. Heavy feeder, making understory difficult.
CHINESE JUJUBE [Zizyphus jujuba]	11m	moderate	minimum	9,000 ppm	Good plant for poor or natural sites. Gives dappled shade.	Withstands drought and poor soil. Full-sun. Survives in exposed locations. Needs training to assume tree proportions.

Shrubs	Height	Growth Rate	Water Requirements	Salinity Tolerance	Uses	Cultivation
BOUGAINVILLEA [Bougainvillea spectabilis]	6m	moderate	moderate	500 ppm	Spectacular garden climber and can be used as ground cover on steep banks.	Needs full-sun and shelter from strong winds. Prefers light, freely draining soil.
FALSE JASMINE [Clerodendron inerme]	3.5m	fast	minimum	8,000 ppm	Can be grown as hedge or groundcover. Can be trained over walls and trellises.	Evergreen. Tolerates drought, exposure, air pollution, and saline conditions. Best in full-sun but endures dappled shade.
HOPSEED BUSH: HOP BUSH [Dodonea viscosa]	3m	fast	minimum	7,000 ppm	Can be used to form a screen. With training it can attain small tree proportions. Does well by the sea.	Evergreen. Tolerates drought, thrives in sun. Excess watering or poor drainage may kill it.
CHINESE HIBISCUS [Hibiscus rosa-sinensis]	3m	moderate	moderate	600 ppm	Ideal border shrub.	Evergreen. Full-sun is ideal, but it will grow in dappled shade. Quickly shows signs of stress if salinity levels are too high.

Table S.4.2   Plant Species Suitable for Sub-tropical Climates

Shrubs (con't)	Height	Growth Rate	Water Requirements	Salinity Tolerance	Uses	Cultivation
LANTANE [Lantana camara]	1.5-3m	moderate	minimum	3,000 ppm	Excellent border plant. Works as good groundcover. Can be trained over walls and trellises.	Evergreen. Prefers full-sun. Straggly if grown in shade. Needs shelter from strong winds.
OLEANDER [Nerium oleander]	6m	fast	minimum	9,000 ppm	Gives good and immediate shelter.	Evergreen. Best in full-sun.
PROSTRATE ROSEMARY [Rosamarinus officinalis]	15-30cm		minimum		Evergreen.	Full sun. Drought-tolerant and suitable for xeriscaping.
YELLOW OLEANDER [Thevetia peruviana]	3m	moderate	moderate	7,000 ppm	Needs adequate space to develop properly.	Evergreen. Latex abundant/poisonous. Prefers full-sun and shelter from strong winds. Rich sandy soil is ideal.

Vines and Groundcovers	Height	Growth Rate	Water Requirements	Salinity Tolerance	Uses	Cultivation
BOUGAINVILLEA [Bougainvillea glabra]	4.7-6m		moderate		Evergreen shrub in frost free locations.	Full-sun.
LANTANE [Lantana camara]	1.8m	fast	minimum		Tolerates salt spray.	Sun to part shade. Very drought resistant. Will adapt to most soil types. Suitable for xeriscaping.

Table S.4.3   Plant Species Suitable for Sub-tropical Climates

FURTHER RESOURCES

Websites:

1. *Colorado WaterWise.* Colorado WaterWise, 2005. Web. 04 June 2010. <http://www.xeriscape.org>.

2. *Irrigation Association.* Irrigation Association, 2010. Web. 04 June 2010. <http://www.irrigation.org>.

3. *Water-Efficient Landscaping.* University of Missouri Extension, 1993. Web. 04 June 2010. <http://muextension.missouri.edu/xplor/agguides/hort/g06912.htm>.

4. United Nations. "Qatar: Country Report to the FAO International Technical Conference on Plant Genetic Resources," *Food and Agriculture Organization of the United Nations.* Food and Agriculture Organization, 1995. Web. 04 June 2010. <http://www.fao.org/ag/AGP/AGPS/Pgrfa/pdf/qatar.pdf>.

5. *The Environmental Institute for Golf.* The Environmental Institute for Golf, 2010. Web. 17 September 2010. <http://www.eifg.org/>.

6. *USGA Course Construction.* United States Golf Association, 2010. Web. 17 September 2010. <http://www.usga.org/course_care/articles/construction/Course-Construction/>.

Publications:

1. State of Qatar. *National Biodiversity Strategy and Action Plan, State of Qatar.* Qatar, 2004. Web. 04 June 2010. <www.cbd.int/doc/world/qa/qa-nbsap-01-en.doc>.

2. United States. Environmental Protection Agency. *Water-Efficient Landscaping: Preventing Pollution & Using Resources Wisely.* Washington: EPA, 2002. Print.

3. Dodson, Ronald G. *Sustainable Golf Courses: A Guide to Environmental Stewardship.* Hoboken: John Wiley & Sons, 2005. Print.

4. Love, Bill. "An Environmental Approach to Golf Course Development." American Society of Golf Course Architects, 2008. Print.

[S.5] Desertification

SCOPE

COMMERCIAL | CORE + SHELL | RESIDENTIAL | EDUCATION | MOSQUES | HOTELS | LIGHT INDUSTRY

- Inherited for SINGLE RESIDENTIAL
- Rated for GROUP RESIDENTIAL

DESCRIPTION

Reverse, prevent, or minimize desertification and protect project site from sandstorms.

GUIDELINES

Desertification is the degradation of land in dry, arid climates. Desertification is caused by a number of human activities and climatic conditions such as overgrazing, irrigation with high saline water, deforestation, extended periods of drought, and wind erosion resulting in infertile and unproductive soil.

Develop plans to minimize, prevent, and reverse desertification through the provisioning of water and the enrichment of soil on the site. Build and modify soil to restore its fertility through the addition of appropriate organic compounds. A soil scientist or specialist can provide recommendations on how to most effectively restore soil productivity by analyzing the existing soil conditions. Establish irrigation measures to provide appropriate levels of water to the site and promote the restoration of vegetation.

Other rehabilitative and preventative measures to minimize desertification and protect buildings from sandstorms include countering erosion through terracing, fixating the soil with protective shelterbelts, windbreaks and sand fences, reintroducing and restoring selected wildlife species on the site, and planting groundcover such as native and/or adapted, non-invasive vines. Plants and trees selected for the landscaping plan should be drought-resistant and have minimal water demand. Refer to the tables in the previous guideline, [S.4] Vegetation, for suggested plant and tree species.

Further desertification can be prevented by integrating land and water management in order to protect soils and vegetation from erosion, salinization, and other forms of degradation.

FURTHER RESOURCES

Websites:

1. United Nations. *United Nations Convention to Combat Desertification.* New York: United Nations Convention to Combat Desertification, 2010. Web. 04 June 2010. <http://www.unccd.int/>.

Publications:

1. United Nations. *Desertification - Coping with Today's Challenges in the Context of the Strategy of the United Nations Convention to Combat Desertification.* Germany: Deutsche Gesellschaft für Technische Zusammenarbeit (GTZ) GmbH, 2008. Print.

[S.6] Rainwater Runoff

SCOPE

COMMERCIAL | CORE + SHELL | RESIDENTIAL | EDUCATION | MOSQUES | HOTELS |
LIGHT INDUSTRY | SPORTS

- Rated for GROUP RESIDENTIAL

DESCRIPTION

Minimize the amount of rainwater exiting the site by collecting or absorbing any rainwater that falls on the
site or building.

GUIDELINES

The Rainwater Management Plan should prevent damage to buildings, hardscapes, and vegetation on
the proposed site and adjacent properties. In cases where the proposed project is sited on previously
developed land, reduce the amount of rainwater runoff through the collection of any excess water on the
site.

Collect and capture rainwater runoff where possible to conserve and reuse the maximum amount of
water. Rainwater that falls on areas of the site that are impervious should be collected, stored, and
treated for reuse. These impervious surfaces may include roads, walkways, parking, and other paved
areas. Harvested rainwater can be used for a multitude of purposes including landscape irrigation, aquifer
recharge, toilet flushing, and janitorial uses.

The volume of rainwater runoff can be further minimized by promoting infiltration and reducing the amount
of impervious surface area on the site. Rainwater that falls directly onto landscaping, including lawn areas
and planting beds, should be directly absorbed. Excess water from rain and irrigation should be captured
using retention basins or allowed to absorb and replenish groundwater resources.

Rainwater that falls on the roof of the building or any other above-ground catchment surfaces, such as
canopies and awnings, should be caught, stored, and reused. Consider permeable shading devices
designed to protect and shade the catchment surface on the roof from solar radiation and to minimize the
loss of water due to evaporation.

Rainwater harvesting, including the collection, storage, treatment, and reuse of rainwater, can be
accomplished using a variety of methods and techniques. Techniques of collection and storage range from
using small rain barrels to underground cisterns depending on specific conditions of the project and site.
Pressure tanks and pumps may be necessary to transfer water between areas of storage, treatment, and
reuse.

Water collected from various harvesting methods can be treated in several ways. Pretreatment often involves the use of screens, filtration devices such as slow sand filters, first-flush diverters, or roof washers to remove sediment and debris from the water before storage. Rainwater collected from the roof and other above-ground surfaces may only require filters to remove dust, while rainwater collected from roads and walkways necessitates more stringent treatment processes to remove oils, fuels, and other harmful materials. Microbiological treatments and disinfection processes are used to remove deleterious substances and may include the use of cartridge filters, ultraviolet light, or membrane filtration. Consider the site, climatic conditions, and all the potential environmental impacts when selecting a rainwater harvesting system and the appropriate filtration and disinfection processes.

FURTHER RESOURCES

Websites:

1. *American Rainwater Catchment Systems Association.* American Rainwater Catchment Systems Association, 1999. Web. 04 June 2010. <http://www.arcsa.org/index.html>.

2. *International Rainwater Catchment Systems Association.* International Rainwater Catchment Systems. Web. 04 June 2010. <http://www.eng.warwick.ac.uk/ircsa/>.

3. *The UK Rainwater Harvesting Association.* UK Rainwater Harvesting Association. Web. 04 June 2010. <http://www.ukrha.org/>.

Publications:

1. Leggett, D., R. Brown, D. Brewer, G. Stanfield and E. Holli. *Rainwater and Greywater Use in Building - Best Practice Guidance C539.* United Kingdom: Construction Industry Research and Information Association, 2001. Print.

[S.7] Heat Island Effect

SCOPE

COMMERCIAL | CORE + SHELL | RESIDENTIAL | EDUCATION | MOSQUES | HOTELS |
LIGHT INDUSTRY | SPORTS

- Rated for GROUP RESIDENTIAL

DESCRIPTION

Minimize heat island effect to reduce impact on the surrounding habitat and environment.

GUIDELINES

Mitigate the impact of heat islands on the environment through the project's design and planning. Consider elements to reduce heat island effect such as the use of vegetation, shading of impervious areas, selection of appropriate building materials, and movement of air on the site.

Select light-colored paving surfaces or use paving materials of low heat capacity to minimize the amount of heat absorption from the sun. In areas with hard, impervious materials, provide shading through the use of vegetation, trees, and architectural features and devices.

Reduce the building footprint to limit the amount of hard surfaces that may absorb heat from the sun. The project layout should use the site in an efficient manner by stacking levels of the building, minimizing parking, and sharing roads and facilities with neighboring properties. Limit the amount of exposed parking pavement on the site by providing well-ventilated underground parking or covering parking with high-reflectance materials or vegetation. Install roofs that are high-albedo or vegetated to reduce heat absorption. Use building materials that are lighter in color in order to reflect the sun's heat rather than absorb it; materials that are high-reflectance or have low solar absorption rates will help to alleviate the thermal environment. However, materials that are too bright and extremely reflective might have a detrimental impact on the environment by creating glare.

Maximize green space and provide an appropriate amount of vegetation and groundcover to help mitigate the effects of heat islands by cooling the air through evapotranspiration and shading the building and pavement on the site. Select native and adapted trees, tall shrubs, and noninvasive vines that have minimal water demand to provide shade on the site. A trade-off exists between the benefit of having abundant vegetation on a site and the amount of water it takes to sustain the vegetation—specify an appropriate level of vegetation as to minimize the demand on the region's limited water resources.

Encourage air movement on the site through the design of the project. Consider the direction of prevailing winds when planning the proposed building placement, orientation, forms, and heights. Ensure continuity between open spaces on the site and provide gaps between buildings in the case of clustered developments in order to encourage airflow.

MOSQUES

The use of fountains and reflecting pools is common in the design of a Mosque and these features can reduce the heat island effect of the site. Water features absorb solar radiation more effectively than many solid building materials and can have a strong impact on passively cooling a space. However, exterior water features should be used sparingly or designed for use with salt water, to reduce the load on fresh water due to the region's high evaporation rate.

SPORTS

Encourage air flow in outdoor or semi-outdoor spaces while limiting the exposure of spectators to extreme solar and heat conditions. Reduce the heat island effect on spectator seating areas by orienting the building to reduce the amount of direct sunlight affecting seating areas. Employ design features, such as shading devices and retractable roofing systems, to reduce negative impacts of solar radiation and the heat island effect. For more information on spectator comfort, see the guidelines for [IE.1] Thermal Comfort.

FURTHER RESOURCES

Websites:

1. United States. Environmental Protection Agency. *Heat Island Effect.* Washington: EPA, 2010. Web. 3 June 2010. <http://www.epa.gov/heatisland/index.htm>.

2. *Heat Island Group.* Lawrence Berkeley National Laboratory, 2000. Web. 3 June 2010. <http://eetd.lbl.gov/HeatIsland/>.

[S.8] Adverse Wind Conditions

SCOPE

COMMERCIAL | CORE + SHELL | RESIDENTIAL | EDUCATION | HOTELS | LIGHT INDUSTRY

- Rated for GROUP RESIDENTIAL

DESCRIPTION

Minimize adverse wind conditions in surrounding spaces at the pedestrian level.

GUIDELINES

Reduce the impact of adverse wind conditions on the proposed site and adjacent properties through the design of the project. Projects should use the Adverse Wind Conditions Calculator and if applicable conduct simulations that take into account the heights and forms of surrounding buildings, speed and direction of prevailing winds, site topography, and relevant landscape features. Adverse wind conditions can be minimized through the appropriate design of the building such as its height and geometry. For instance, the form of the roof can be designed to lead winds away from the ground level of the building. Harmful wind conditions can also be prevented using elements such as fences and barrier trees and plants.

Ensure that entrances to the proposed buildings are not exposed to excessive wind conditions by considering their orientation on the site and utilizing screening systems. Screening strategies may include architectural devices, or natural buffers such as trees and plants. In the case of clustered developments, consider the distribution and orientation of individual buildings in regards to wind movement on the site.

[S.9] Noise Pollution

SCOPE

COMMERCIAL | CORE + SHELL | EDUCATION | MOSQUES | HOTELS | LIGHT INDUSTRY | SPORTS

DESCRIPTION

Minimize the level of noise produced by the project affecting nearby buildings and the surrounding environment.

GUIDELINES

Noise-generating facilities or functions from the proposed building or development should not negatively impact the surroundings, including wildlife habitats and noise-sensitive buildings on-site or on adjacent sites. Minimizing the amount of generated noise, placing the noise-generating facility or equipment in a suitable location on the site, and applying measures to prevent the dispersion of noise are some strategies that help mitigate noise pollution. The sources of noise generated by buildings on the site is generally due to air conditioning systems (HVAC), laundry facilities, the activities of building users, wind-generated noise, and the sounds emanating from parking facilities and roads.

Continuously operating mechanical systems in buildings can constantly affect surrounding indoor and outdoor environments. To effectively maintain noise levels within these environments, establish noise thresholds so that mechanical operations will not exceed comfortable noise levels.

Locate noise-generating facilities underground or away from the periphery of the site, especially if adjacent properties include noise-sensitive areas. Noise-sensitive areas include sites or buildings where the users are likely to be negatively affected by the noise generated by the new building or development. These may include residential areas, medical facilities, libraries, classrooms, places of worship, wildlife areas, parks, or gardens. Prevent the dispersion of noise to adjacent properties using noise control measures such as sound-baffling walls, fences, or trees to minimize the effects of noise pollution. Through the design of the project site plan, isolate or separate building spaces that will contain noise-generating activities from any noise-sensitive areas adjacent to the site.

Consider the form and arrangement of proposed buildings when trying to reduce noise pollution generated from wind conditions on the site. The building's exterior elements can also generate noise from air movement; the form, material, and positioning of roofs, wall systems, louvers, and other exterior building elements may produce wind-generated noise and impact noise-sensitive areas within the new project and in the larger community.

Minimize the noise produced by on-site traffic through the careful placement of roads and parking locations on the site. Avoid placing plazas, courtyards, pedestrian walkways, recreation areas, roads, and parking facilities near noise-sensitive areas in the development, and use sound barriers and buffers where necessary.

HOTELS

During nighttime operations, noise levels should be lowered to ensure comfort levels for hotel guests.

SPORTS

Events held at sport facilities will usually significantly exceed the noise levels generated during non-event periods. This increase in noise comes primarily from two sources: the crowd and the sound system. In order to ensure that the noise level from the sound system does not exceed noise pollution limits, or to improve the score a facility receives for peak events, an electronic cap should be placed on the output of the sound system. Even if the majority of the noise generated during an event will not exceed this limit, having an electronic limit on the sound system will prevent the occasions where the noise generated will become a nuisance.

The noise generated by a crowd during a peak event is far more difficult to control, but it can be managed. The score received for the Noise Pollution criteria is based on the average noise generated by all events expected to be held in a given year. The average may be affected by reducing the number of events that are expected to be loud in comparison to the number of events that are expected to be quieter. If a local stadium is expected to host many quiet sporting events and several loud rock concerts each year, then reducing the number of concerts compared to the number of games will improve the score. Additional means of lowering crowd noise can be achieved through consultation with an acoustical engineer who can advise on how to construct a stadium that will leak minimal noise levels from a crowd. Also, banning or limiting noise making devices that are popular at many sporting events, such as the vuvuzela, which caused problems during the 2010 World Cup in South Africa, will greatly reduce unwanted crowd noise.

The impact of noise from events should be minimized by scheduling the events to occur during the least sensitive time periods. While the score received in the Noise Pollution criterion is not time dependent, events should be scheduled so that they occur when minimal impact will be felt by the surrounding facilities and homes. If the sports facility is located near residences, noisy events should be avoided later in the evening hours while sports facilities near office buildings or businesses would be better off avoiding noisy events during the day.

FURTHER RESOURCES

Publications:

1. *Acoustics - Description, measurement and assessment of environmental noise - Part 1: Basic quantities and assessment procedures.* ISO 1996-1:2003. International Organization for Standardization, 2003. Web. 30 June 2010. <http://www.iso.org/iso/iso_catalogue.htm>.

2. *Acoustics - Description, measurement and assessment of environmental noise - Part 2: Determination of environmental noise levels.* ISO 1996-2:2007. International Organization for Standardization, 2007. Web. 30 June 2010. <http://www.iso.org/iso/iso_catalogue.htm>.

3. *Acoustics - Description and measurement of environmental noise - Part 3: Application to noise limits.* ISO 1996-3:1987. International Organization for Standardization, 1987. Web. 30 June 2010. <http://www.iso.org/iso/iso_catalogue.htm>.

[S.10] Light Pollution

SCOPE

COMMERCIAL | CORE + SHELL | EDUCATION | HOTELS | LIGHT INDUSTRY | SPORTS

DESCRIPTION

Minimize the amount of light emitted to the exterior from the building or site.

GUIDELINES

Reduce the effects of light pollution that result from external lighting and light trespass from within the building. Light pollution can impact the environment through effects, such as glare and sky glow, which should be minimized through the design of an efficient and sustainable lighting scheme. Consider consulting a lighting professional to provide lighting recommendations based on the specific project needs, functions, and site conditions, with the intention of reducing the effects of light pollution. Computer modeling and simulations can be used to analyze lighting conditions and thereby improve the lighting design of the project.

Determine which zone (Zone E1 – Zone E4) the proposed project is classified under according to the IESNA-RP 33 standards. Comply with the requirements and standards for that specific zone. Lighting ordinances or bylaws relevant to the site should be reviewed and considered before designing the lighting scheme. Where possible, minimize lighting on the project site including lighting used for architectural and landscape features such as plazas, courtyards, walkways, buildings, trees, and planting beds. Where lighting is used to highlight specific features, cluster the fixtures in the necessary areas to reduce light dispersion and disturbance to the environment. If external lighting is needed for safety and security purposes, avoid upward lighting schemes, and consider the use of low-intensity, shielded fixtures; highly focused, low-voltage lamps; full cutoff luminaires; low-reflectance surfaces; and low-angle spotlights to minimize light pollution. All non-essential external lighting should be automatically switched to lower illumination levels or turned off after normal operation hours to minimize energy use and environmental disturbance. Lighting in parking lots should be zoned so that the area closest to the building can be activated without illuminating the entire lot.

All selected lighting fixtures and equipment should meet high-performance and quality standards, in order to maximize visual quality, provide efficient and effective lighting, and reduce the need for maintenance. Use energy efficient fixtures to reduce lighting power and illumination intensity.

Light trespass occurs when direct beams from interior light fixtures permeate transparent and translucent surfaces in the building envelope and negatively affect the exterior site. Determine the direction and intensity of each fixture type, and locate interior lighting fixtures so that the direct beam illumination intersects opaque building surfaces or dissipates before reaching the exterior. The direction of maximum luminous intensity can be found using manufacturers' candela plots or photometric data. The lighting system should allow for the automatic switch-off of all non-emergency interior lighting fixtures after normal hours of operation. Consider using devices such as automatic sweep timers, occupancy sensors, motion detectors, or programmed lighting control panels. The lighting system design must include the capability for manual override for use after normal occupancy hours.

EDUCATION | SPORTS

Nighttime, outdoor sporting events require intense illumination. Use shielded luminaires on lighting fixtures at athletic facilities to significantly reduce light trespass to adjacent sites. Luminaries for athletic facilities that are adjacent to roadways should not create any glare on the road.

Since a variety of user groups visit sports and educational facilities, including full-time occupants such as employees and students and part-time occupants who only attend events, all external and internal lighting should be zoned to meet the demands of these user groups and improve efficiency. In cases where a smaller number of occupants use the facility, fewer lights may be turned on, whereas peak event times may require maximum lighting use.

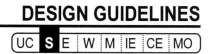
FURTHER RESOURCES

Websites:

1. *Institution of Lighting Engineers.* Institution of Lighting Engineers. Web. 10 August 2010. <www.ile. org.uk>.

2. *Illuminating Engineering Society.* Illuminating Engineering Society of North America, 2010. Web. 10 August 2010. <http://www.ies.org/>.

3. *International Dark-Sky Association.* IDA. Web. 10 August 2010. <http://www.darksky.org/mc/page. do>.

Publications:

1. *Lighting for Exterior Environments.* IESNA RP-33-99. Illuminating Engineering Society of North America, 1999. Print.

2. *Roadway Lighting ANSI Approved.* ANSI/IESNA RP-8-00. American National Standards Institute/ Illuminating Engineering Society of North America, 2000. Print.

3. Chartered Institution of Building Services Engineers. *The Outdoor Environment, Lighting Guide 6.* CIBSE, 1992. Print.

4. The Institution of Lighting Engineers. *Guidance Notes for the Reduction of Obtrusive Light, GN01.* ILE, 2005. Print.

5. The Institution of Lighting Engineers. *Guidance Notes for the Reduction of Light Pollution.* ILE, 2000. Print.

6. Commission Internationale de L'Eclairage. *Guide on the Limitation of the Effects of Obtrusive Light from Outdoor Lighting Installation, CIE 150:2003.* CIE, 2003. Print.

7. Institution of Lighting Engineers. *The Brightness of Illuminated Advertisements, Technical Report No.5.* ILE, 2001. Print.

8. British Standards Institute. *Code of Practice for the Design of Road Lighting – Lighting of Roads and Public Amenity Areas.* BS5489-1. British Standards Institute, 2003. Print.

[S.11] Shading of Adjacent Properties

SCOPE

COMMERCIAL | CORE + SHELL | RESIDENTIAL | EDUCATION | MOSQUES | HOTELS | LIGHT INDUSTRY | SPORTS

- Rated for GROUP RESIDENTIAL

DESCRIPTION

Minimize the shading of adjacent properties to allow for access to daylight or solar energy potential.

GUIDELINES

Ensure that all existing and future buildings on adjacent sites have direct access to daylight, so as not to reduce the potential benefits of solar access and energy. Limit the shade from the proposed building that may be cast upon the southern surfaces of buildings immediately to the north of the site. Conduct simulations to determine the amount of shade and shadows that will fall on building surfaces of adjacent sites, and adjust the proposed building's location, orientation, and height to minimize obstruction of daylight. Daylight obstruction to nearby sites can also reduce the ability for vegetation and landscaped areas to access necessary sunlight.

While efforts should be made not to shade adjacent properties with the proposed building, buildings within clustered developments can be designed to shade one another in order to reduce heat gain from the sun within the proposed development.

[S.12] Parking Footprint

SCOPE

COMMERCIAL | CORE + SHELL | RESIDENTIAL | EDUCATION | MOSQUES | HOTELS | LIGHT INDUSTRY | SPORTS

• Rated for GROUP RESIDENTIAL

DESCRIPTION

Minimize the parking footprint within the development to reduce the amount of impermeable surfaces on the site.

GUIDELINES

Implementing efficient parking strategies early in the design process can minimize the on-site parking footprint while adequately serving the needs of new development. Parking demand for a building or development should be based on its use and intensity. The industry standard methodology for parking distribution is the Institute of Transportation Engineers (ITE) *Parking Generation Manual;* however, ITE standards are based on observations of peak demand for parking at single-use developments in relatively low-density settings with little public transit and tend to overestimate actual needs—especially in terms of creating a sustainable parking scheme.

Minimum parking requirements for new construction are usually established by international standards, and the requirements vary based on typology with retail requiring the most parking. Consider lowering or abolishing minimum parking requirements for the building, and establish maximum parking requirements based on usage, transit options, demographics, and density. Higher density areas allow for reduced auto ownership because destinations are closer together, and more places can be reached on foot and by bicycle. In addition, employ transit-oriented development strategies that concentrate higher densities around public transportation nodes.

Develop a Parking Footprint Plan, and consider parking strategies that can shrink the overall parking footprint and provide flexibility to meet always changing demand. Structured and underground parking minimizes the size and visibility of parking areas while increasing the walkability and vibrancy of the site. Use shared parking areas for adjacent facilities to reduce the number of needed curb cuts and enhance pedestrian access. Parking may also be shared in areas that combine different zoning types, such as office and residential, where parking spaces may be used at different times of day. On-street parking not only reduces construction costs and parking footprint, but it also helps to slow traffic and shield pedestrians.

The heat island effect and rainwater runoff may be managed through the use of high-albedo or pervious paving systems. Ultimately, access drives, internal circulation drives, parking areas, and walkways should be designed to provide safety, convenience, and accessibility for motorists as well as pedestrians.

FURTHER RESOURCES

Websites:

1. United States. Environmental Protection Agency. *Urban Sustainability & the Built Environment.* 02 February 2010. Web. 11 June 2010. <http://www.epa.gov/sustainability/builtenvironment.htm>.

2. *Institute of Transportation Engineers.* Institute of Transportation Engineers, 2010. Web. 11 June 2010. <http://www.ite.org/>.

3. Smart Growth Online. Web. 11 June 2010. <http://www.smartgrowth.org/>.

4. *Urban Land Institute.* Urban Land Institute, 2010. Web. 11 June 2010. <http://www.uli.org/>.

5. *Online Transportation Demand Management Encyclopedia.* Victoria Transportation Policy Institute, 27 February 2010. Web. 28 June 2010. <http://www.vtpi.org/tdm/tdm12.htm>.

Publications:

1. United States. Environmental Protection Agency. *Parking Spaces / Community Places: Finding the Balance through Smart Growth Solutions.* Washington: EPA, 2006. Print.

2. *Shared Parking, 2nd Edition.* Urban Land Institute: Washington, 2005. Print.

3. *Parking Generation, 3rd Edition.* Institute of Transportation Engineers: Washington, 2002. Print.

4. *Traffic Engineering Handbook, 6th Edition.* Washington: Institute of Transportation Engineers, 2009. Print.

5. *Transportation and Land Development.* Washington: Institute of Transportation Engineers, 2002. Print.

6. Shoup, Donald. "The Trouble with Minimum Parking Requirements." *Transportation Research Part A.* Vol. 33. Victoria: Victoria Transport Policy Institute, 1999. 549-574. Print.

7. "Truth in Transportation Planning." *Transportation Research Board 80th Annual Meeting.* Washington: Transportation Research Board, 2001. 1-15. Print.

8. Singelis, Nikos, Lisa Nisenson and Martina Frey. "Lots and Lots of Parking Lots." *Stormwater Solutions.* Chicago: Scranton Gillette Communications, Jan. 2008. 10-13. Print.

[S.13] Shading

SCOPE

COMMERCIAL | CORE + SHELL | RESIDENTIAL | EDUCATION | MOSQUES | HOTELS | LIGHT INDUSTRY | SPORTS

- Rated for GROUP RESIDENTIAL
- Inherited for SINGLE RESIDENTIAL

DESCRIPTION

Encourage shading for areas of the site not occupied by buildings or roadways.

GUIDELINES

Public spaces are most often used when thermal conditions are closest to an individual's thermal comfort range. Due to the hot temperatures and intense solar radiation in Qatar, any areas of the proposed site that are not occupied by buildings and roads should be vegetated and shaded to the greatest extent possible. Utilize shading devices and/or trees over sidewalks, walkways, and bicycle paths to protect pedestrians and cyclists from the sun. Provide shaded paths and walkways between adjacent buildings, parking, green spaces, and other shared facilities. Open spaces can be protected from direct sunlight using various architectural and landscaping methods. For example, shading devices, such as trellises, awnings, canopies, or built structures, can be used to create shade. Trees, shrubs, and other forms of vegetation can also provide shade to the site.

Picnic and seating areas may benefit from permanent shading structures, such as pavilions or fabric canopies, that provide full-time coverage. When designing outdoor shelters and shaded areas, projects should consider the time of day when the spaces are used in order to determine proper shading coverage. Consider the durability of materials used for architectural shading devices, as well as all operational, maintenance, and safety issues.

Shading strategies can be derived from studying the sun angles on the site and using sun path diagrams and computer simulations. Use computer simulations when designing a sheltered area to determine the influence of the time of day, location, and orientation on thermal comfort. Providing proper shading for entrances, parking lots, pedestrian pathways, picnic areas, and other common areas helps to reduce heat gain, mitigate heat island effects, and encourage pedestrian activity on the site.

RESIDENTIAL

Encourage publicly accessible green spaces within residential developments by providing vegetated and shaded areas to promote community development and recreation. These spaces allow for outdoor activity, community interactions, and educational spaces improving the health and well-being of the development and/or district. These spaces provide ecological benefits such as reduced heat island effects, greater stormwater control, and wildlife habitats.

FURTHER RESOURCES

Publications:

1. Carmona, Matthew et al. *Public Places - Urban Spaces: The Dimensions of Urban Design.* Oxford: Elsevier, 2003. Print.

2. Moll, Gary and Sara Ebenreck. *Shading Our Cities: A Resource Guide for Urban and Community Forests.* Washington: American Forestry Association, 1989. Print.

[S.14] Illumination

SCOPE

COMMERCIAL | CORE + SHELL | RESIDENTIAL | EDUCATION | MOSQUES | HOTELS | LIGHT INDUSTRY | SPORTS

- Rated for GROUP RESIDENTIAL

DESCRIPTION

Ensure light levels have been designed in line with best practices for visual performance and comfort.

GUIDELINES

Design the project's lighting systems to ensure adequate and uniform illumination levels for the visual comfort and well-being of the project users. Provide light levels in accordance with those recommended in the *IESNA Lighting Handbook, 9th Edition*. Ensure that they do not drastically exceed the recommended maximum illumination levels. Spaces that are well lit will help maintain visual comfort and deter crime or vandalism.

Identify where and when lighting is needed. Choose the most efficient light source that meets the task requirement. Minimize illumination intensity and energy use through the specification and selection of energy efficient fixtures that use efficacious sources. Shielded fixtures and full cut-off fixtures with efficient light bulbs are more cost-effective because they use less energy by directing the light toward the ground. Additionally, these types of fixtures reduce light trespass and glare. Contain light within the site by carefully selecting, locating, mounting, and aiming the luminaires to reduce light pollution and make the system efficient.

In addition to the lateral positioning, type, and the distribution of the luminaires, projects should design the park illumination system with consideration for the mounting height in conjunction with spacing. Coordinate light fixture layouts in conjunction with benches and picnic areas to maximize lighting efficiency. Take the location of trees and shrub growth into account when locating lighting fixtures, as they may prevent light from reaching the intended area. Consider the paving material and differences in the reflectance characteristics when calculating light levels.

In a public setting, facial identification is an important element for security reasons. The following factors should be considered when specifying fixtures to improve light quality and reduce shadows: luminance ratio limits, veiling reflections, reflected glare, shadows, color, and intensity. The project should establish a periodic relamping and cleaning program to maintain the required lighting levels. Easy access to light fixtures should be considered when selecting the fixture type and location.

SPORTS

Sports facilities tend to admit and release large volumes of pedestrian traffic as events begin and end. Therefore, all entry points and egresses should be well lit to allow spectators to enter the stadium or return to their transportation with ease and reduce congestion at entry and exit points. This will create a safer and more pleasant environment for spectators.

FURTHER RESOURCES

Websites:

1. *Illuminating Engineering Society.* Illuminating Engineering Society of North America, 2010. Web. 10 August 2010. <http://www.ies.org/>.

2. *International Dark-Sky Association.* IDA. Web. 10 August 2010. <http://www.darksky.org/mc/page.do>.

Publications:

1. *Recommended Practice of Daylighting, RP-5-99.* New York: Illuminating Engineering Society of North America, 1999. Print.

2. Rea, Mark S., ed. *The IESNA Lighting Handbook.* 9th ed. New York: Illuminating Engineering Society of North America, 2000. Print.

3. *Code for Lighting: Part 2.* United Kingdom: Chartered Institution of Building Services Engineers, 2004. Print.

4. Boed, Viktor. *Controls and Automation for Facilities Managers: Applications Engineering.* Boca Raton: CRC Press, 1998. Print.

[S.15] Pathways

SCOPE

COMMERCIAL | CORE + SHELL | RESIDENTIAL | EDUCATION | MOSQUES | HOTELS |
LIGHT INDUSTRY | SPORTS

- Rated for GROUP RESIDENTIAL

DESCRIPTION

Encourage sustainable development with efficient, user-friendly pedestrian pathways by ensuring that all pathways leading through the site feature adequate signage and accommodate all users, including those with disabilities.

GUIDELINES

Design pathways to provide direct and safe connections for pedestrians and bicyclists to maximize convenience to building occupants and other users of the site. If possible, design pathways to be visible from other areas on the site to foster a sense of security and promote a safer environment for pedestrians. Additionally, plan pathways to avoid loading zones, mechanical equipment, and other spaces unpleasant for pedestrians.

Sites that are easy to navigate enhance users' sense of safety, minimize their anxiety, and improve their environmental awareness. Create an environment that makes it easy and intuitive for all users, including those with disabilities, to orient themselves and navigate from place to place. In this criterion, accessibility refers to the physical access for users with disabilities. Therefore, accessible pathways refers to pathways that are designed to be used by users with disabilities.

Create a network of pathways, including pedestrian trails, bicycle paths, and accessible paved pathways, on the project site. Consider existing pathways on adjacent sites when designing the layout of the proposed project to ensure appropriate and efficient connectivity for pedestrians, bicyclists, and users with disabilities. Pathways should be designed to provide direct and safe connections. Building frontages and entrances should face the street to promote active pathways and streetscapes.

All pathway surfaces should be solid, smooth, free of obstruction, and able to withstand inclement weather regardless of the specific material used. Provide sufficient space for the anticipated foot traffic, as well as the presence of other features on the pathways. Separate pathways from roadways with the use of raised sidewalks, curbs, or bollards to clearly identify the paths.

Accessible Pathways

Provide accessible pathways, including sidewalks alongside roadways and accessible paved pathways between all building entrances, throughout the project site. Paved pathways meant for users with disabilities should meet the requirements of the proposed standards by *Architectural and Transportation Barriers Compliance Board; Architectural Barriers Act (ABA) Accessibility Guidelines for Outdoor Developed Areas*. Provide accessible pathways with the appropriate surface materials, proper width and slope, and the appropriate lighting scheme to ensure user comfort and safety. At instances where the path is not wide enough, provide passing spaces, as recommended by the accessibility guidelines, to accommodate users with disabilities. Consider the use of benches and other programmed elements, such as shaded accessible rest areas, where necessary.

Signage & Wayfinding

Use design features, including signage, awnings, and prominent entrance areas, to assist with wayfinding and walkability. Pathways and building entrances should be clearly labeled in order to allow for convenient wayfinding between facilities within the site, as well as to adjacent properties and public transportation nodes. When vehicular traffic is anticipated to cross pathways, additional street markings and signage should be used to clearly mark traffic patterns. Place signage at regular intervals along pathways in a position where it is visible to all the intended users. Provide vehicular directional signage, pedestrian directional signage, labels for pathways, safety and advisory warnings, and accessible signage for pathways intended for use by users with disabilities.

SPORTS

Sports facilities tend to admit and release large volumes of pedestrian traffic as events begin and end. Pathways between buildings, public transit, and parking areas should be designed to encourage the rapid and orderly filling and emptying of the facility. Allowing multiple entry points and egresses supported by well-lit exterior pathways will allow spectators to enter the stadium or return to their transportation with ease and reduce congestion at entry and exit points. This will create a safer and more pleasant environment for spectators.

Alternative pathways should be provided for a variety of backstage activities that will allow for the separation of players and/or performers from the general spectators. This will allow for greater ease of flow in pedestrian traffic while entering and exiting the facility by preventing gawking, photo taking, or other idling. Separating the players and/or performers from the spectators is also a safety concern that must be addressed. This becomes less important for facilities that do not host events with well-known players and/or performers such as high school or small college sports facilities.

DESIGN GUIDELINES

FURTHER RESOURCES

Publications:

1. United States. U.S. Department of Transportation. Axelson, Peter, et al. *Designing Sidewalks and Trails for Access: Part I of II: Review of Existing Guidelines and Practices.* Washington: U.S. Department of Transportation, Federal Highway Administration, 1999. Web. 29 June 2010. <http://www.fhwa.dot.gov/environment/sidewalks/index.htm>.

2. Minnesota. Access Minneapolis. *Design Guidelines for Streets and Sidewalks.* Minneapolis: City of Minneapolis Public Works, 2008. Web. 29 June 2010. <http://www.ci.minneapolis.mn.us/public-works/trans-plan/DesignGuidelines.asp>.

3. Architectural and Transportation Barriers Compliance Board. Architectural Barriers Act (ABA) Accessibility Guidelines for Outdoor Developed Areas; Draft Final, 2009. <http://www.access-board.gov/outdoor/draft-final.htm#text>

4. National Center on Accessibility. *"What is an Accessible Trail?"* Bloomington, IN: National Center on Accessibility, Indiana University-Bloomington, 2002. Web. 29 June 2010. <http://www.ncaonline.org/?q=node/659>

[S.16] Mixed Use

SCOPE

COMMERCIAL | CORE + SHELL

DESCRIPTION

Maximize the number of major uses within the project to reduce the need for transport.

GUIDELINES

Consider a range of major uses within the project site to promote an active streetscape and minimize the need for transportation. Active streetscapes create safer environments and communities for building occupants and pedestrians. For example, a building that contains offices as well as restaurants will support an active streetscape because the building will be regularly utilized after office hours. Providing for many services and uses within the same building or development will reduce the need for transportation and encourage pedestrian activity, thereby minimizing impact to the environment through vehicular emissions. Additionally, consolidating uses and shared services can reduce the building/development footprint and minimize the impact on the environment.

If planning for multiple major uses, the project should also consider the impact of traffic loads and on-site parking facilities on surrounding streets and intersections. Analyze the site and its relationship to adjacent properties to anticipate congestion and traffic issues that may arise from the new development, especially if the project will contain a diversity of services and attract a large number of users. Ensure that various uses within the same building or development will not detract from one another or cause unhealthy competition.

ENERGY [E]

The Energy Guidelines are a series of recommendations for designers to improve building energy performance in Qatar. This standard is referenced by the Global Sustainability Assessment System (GSAS) and at some later stage by the separate Qatar Building Energy Code that will impose limits on the use of energy of buildings; thus, contributing to a less energy intensive economy and a more sustainable environment.

The Energy Guidelines present the components and parameters that affect the total building energy consumption. They range from building envelope elements to HVAC system components. None of these factors can be looked at in isolation; only the calculation of the overall outcome will show the effectiveness of the design.

GSAS introduces the Energy Performance Coefficient or EPC. It is a quantified measure for understanding how well a building design performs in terms of energy use compared to a baseline design. EPCs have been introduced at three levels of design: the building, its systems, and its supply network. This leads to the definition of EPC_{nd} (energy demand performance), EPC_{del} (delivered energy performance), and EPC_p (primary energy performance).

EPC_{nd} is a measure for comparing the effectiveness of the site selection and major building and envelope parameters in removing internal loads and shielding against the outside environment, while maintaining the required indoor comfort.

EPC_{del} is a measure for comparing the effectiveness of the designed building systems in meeting the energy needs of the building.

EPC_p is a measure for comparing different energy delivery systems on and to the site taking the supply network into account.

All EPC measures are based on normative, standardized calculations of outcomes (energy use as designed), divided by a reference value for a given building type in Qatar. The Qatar Building Energy Code will deem a proposed design acceptable if each level the EPC value is better than a defined threshold. Analogously, in the GSAS scoring method, the lower the energy performance (the lower the EPC value), the higher the resulting GSAS score.

Building energy demand on subsystems, such as HVAC, lighting, and power, can be reduced by integrated design that minimizes building loads by selecting an appropriate site and making good selections in building system types and their sizing. These guidelines provide recommendations for design teams to meet the imposed limits on energy use with pointers on the standard calculations that have to be conducted to demonstrate compliance. The most important design parameters of buildings that should be addressed are the following:

1. Building Site Selection:
   - Site Location
   - Site Orientation

2. Building Envelope:
    - Building Footprint
    - Roof
    - Walls
    - Floors
    - Slabs
    - Doors
    - Windows
    - Skylights

3. Building Internal Load:
    - Lighting Load
    - Appliance Load
    - Occupancy Load

4. HVAC Systems and Equipment
    - Cooling System
    - Chillers
    - Cooling Towers
    - Chilled Water Pumps
    - Condenser Water Pumps
    - Heat Exchangers
    - Pipes, Connections and Valves
    - Condensing Units
    - Cooling system insulation

5. Heating System*:
    - Boilers
    - Heat Exchangers
    - Hot Water Pumps
    - Pipes, Connections and Valves
    - Electric Heaters
    - Heat Pumps
    - Heating system insulation

    *Note: In most building types in Qatar, a heating system will not be present.

6. Ventilation System:
    - Air Handling Units
    - Fan Coil Units
    - Intake & Exhaust Fans
    - Ducts, Outlets, Connections and Dampers
    - Air system insulation

7. Controls:
   - Control System
   - Control Strategies

8. Renewable Strategies:
   - Photovoltaic Cells
   - Solar Panels
   - Ground source Heat Pumps
   - Wind
   - River/Ocean

9. Commissioning:
   - Preconstruction Commissioning

Making the correct decisions about the above elements will result in a more sustainable building that consumes less energy. Additionally, factors that could mitigate environmental impacts due to energy use include: designing the building to lower its energy demand; selecting efficient building systems; lowering the demand on nonrenewable sources of energy, thereby reducing harmful emissions and depletion of fossil fuels; and minimizing the amount of harmful substances produced by the energy delivery systems and the energy supply network. In the following sections, techniques and considerations for making these choices will be provided.

Criteria in this category include:

E.1	Energy Demand Performance
E.2	Energy Delivery Performance
E.3	Fossil Fuel Conservation
E.4	CO_2 Emissions
E.5	NO_x, SO_x, & Particulate Matter

[E.1] Energy Demand Performance

SCOPE

COMMERCIAL | CORE + SHELL | RESIDENTIAL | EDUCATION | MOSQUES | HOTELS |
LIGHT INDUSTRY | SPORTS

- Rated for SINGLE and GROUP RESIDENTIAL

DESCRIPTION

Establish energy demand performance levels for the building in order to reduce environmental and economic impacts associated with excessive energy use.

GUIDELINES

EPC_{nd} is responsive to changes in building site, building envelope, and internal loads in the building. Improving any of these factors will have a positive impact on EPC_{nd}.

Major attention should be given to the shell such as the placement of windows and the choice of U-Values. The choice of shading, solar reflection, and other measures to reduce the Solar Heat Gain Factors have the greatest effect on reducing the predominant energy consumption for cooling.

Another major factor is proper lighting design. Using strategies that implement use of daylighting for large parts of the building is highly recommended. This can be accomplished in slender buildings with lower interior partitions or with vision windows between the exterior and interior zones. This can greatly reduce the required general lighting in the building; thus, reducing electricity for lighting as well as reducing cooling loads. The use of efficient lighting fixtures, and in the future, solid state lighting is also highly recommended.

Building Site Selection

1. Select a site where the adjacent buildings or landmarks can provide shading without negatively affecting access to daylight.

2. The orientation of a building is important, it determines the duration of sun exposure on windows and other surfaces of the building. Figure 1 shows that in Qatar, the optimum orientation is close to but slightly East of due South, centered at 172.5°, clockwise from North.

Optimum Orientation
Location: Doha, Qatar

Orientation based on the average daily
incident radiation on a vertical surface.
Underheated Stress: 0.0
Overheated Stress: 2207.3
Compromise: 177.5°

Avg. Daily Radiation at 176.0°
Entire Year: 1.92 kWh/m2
Underheated: 3.51 kWh/m2
Overheated 0.31 kWh/m2

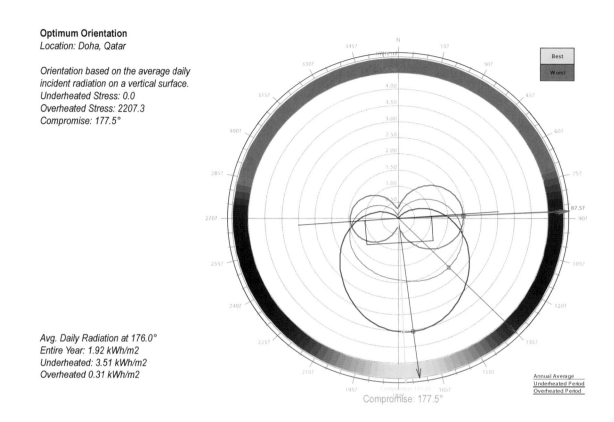

Figure 1   Optimum Building Orientation in Qatar

Building Envelope

1. Specify envelope elements with low U-Values (high R-Values) to reduce both solar and conductive heat gains and losses. Baseline recommendations for selecting the U-Values of envelope elements can be found in *ASHRAE 90.1* for different climatic zones. Reference Figure 2 below for an example of heat transfer through a typical wall.

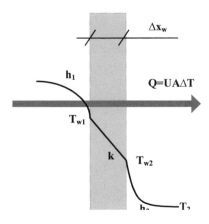

Figure 2   Heat Transfer Through Typical Wall

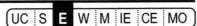
2. Specify all windows and skylights with low Solar Transmittance to control solar gain and reduce cooling load. Baseline recommendations for selecting the Shading Coefficient of windows and skylights can be found in *ASHRAE 90.1* for different climatic zones, including zone 1 for Qatar.

3. Design a building with a minimum amount of windows on the west, east, and south orientations.

4. Design a building with a maximum amount of windows on the North orientation to benefit from indirect daylighting. Reference Figure 3 below for an example of top and side daylighting.

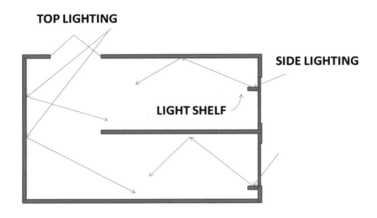

TOP & SIDE DAYLIGHTING

Figure 3   Top & Side Daylighting

5. Design the building with the minimum acceptable quantity of windows, especially those oriented towards the west.

6. The following charts are tailored specifically for Qatar, and show the effects of total monthly transmission and solar heat gains on one square meter of opaque or glazing surface for the four orientations. The charts depict the effect of installing two different building materials (different U-value and solar transmittance).

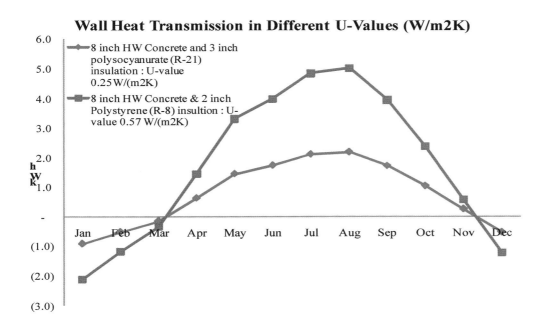

Figure 4   Wall Heat Transmission in Different U-Values (W/m2K)

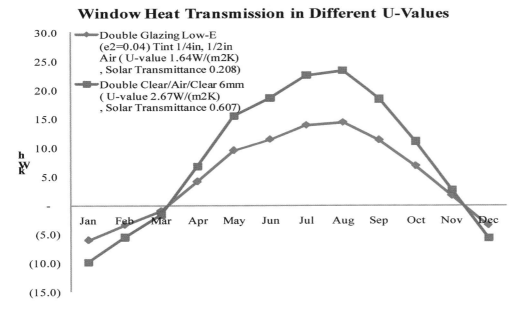

Figure 5   Window Heat Transmission in Different U-Values (W/m2K)

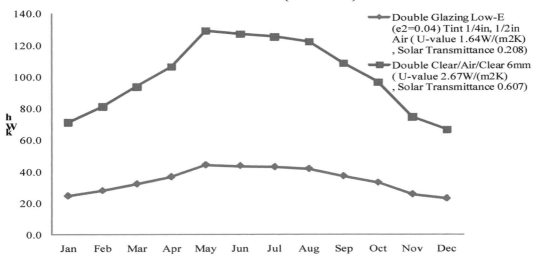

Figure 6   Horizontal Solar Heat Gain in Different Solar Transmittance (kWh/M2)

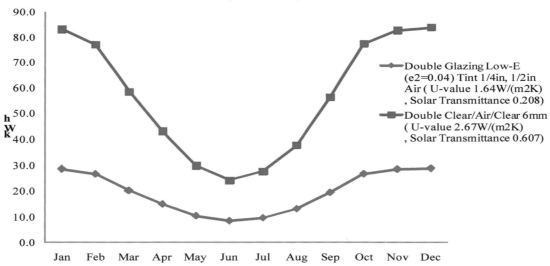

Figure 7   South Solar Heat Gain in Different Solar Transmittance (kWh/m2)

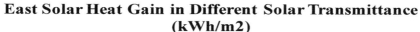
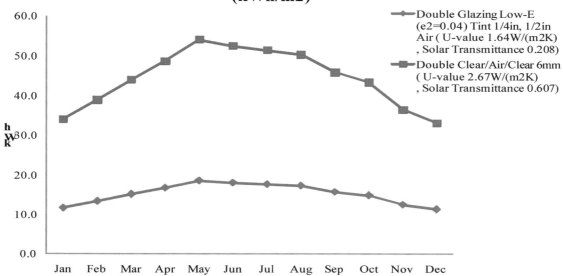

Figure 8   East Solar Heat Gain in Different Solar Transmittance (kWh/m2)

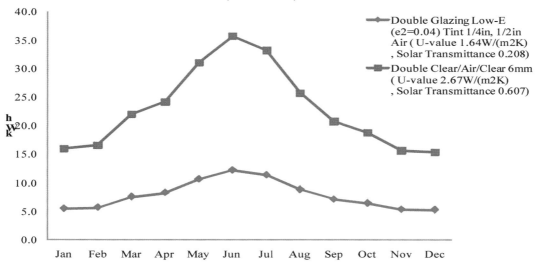

Figure 9   North Solar Heat Gain in Different Solar Transmittance (kWh/m2)

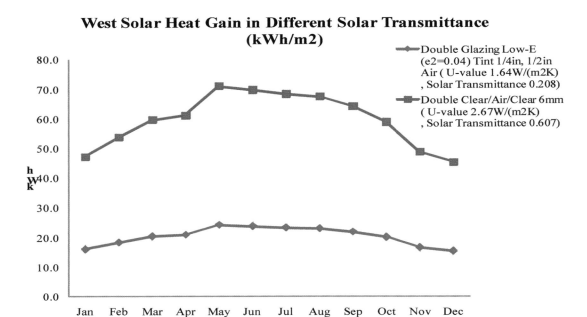

Figure 10   West Solar Heat Gain in Different Solar Transmittance (kWh/m2)

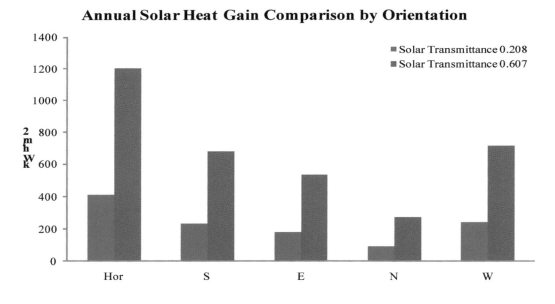

Figure 11   Annual Solar Heat Gain Comparison by Orientation (kWh/m2)

7. Utilize passive solar design considerations. Reference Figure 12 below for an example of passive solar design.

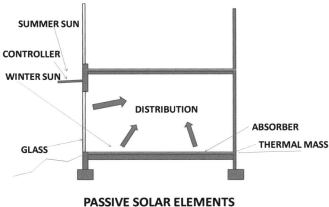

PASSIVE SOLAR ELEMENTS

Figure 12   Example of Passive Solar Design

8. Increase roof surface reflectance and emittance through the use of reflective paints, materials or coatings.

9. To reduce the chance of introducing uncontrolled and unconditioned outside air into the building, do not specify louvers on the side of the building facing high wind pressure.

10. Specify all the joints around the windows, skylights, and doors, as well as junctions between walls and other structural elements, to be built airtight to minimize air leakage rates.

11. Specify all the walls, floors, and chase penetrations with pipes and ducts to be filled with the proper material. Proper material specification will minimize the rate of air leakage occurring between the pipe/ duct exterior and the penetration opening.

12. Specify vapor retardant on all the exterior structural elements to prevent humidity from entering or leaving the building.

13. Use hybrid ventilation strategies, such as providing operable windows, where possible. If such strategies are used, pay particular attention to temperature and humidity control in the conditioned spaces. The entry of unconditioned and uncontrolled outside air into the building should be monitored and system provisions should be made to safeguard against inefficiencies and condensation. A thorough analysis shall be conducted before specifying operable or non-operable windows or other hybrid ventilation strategies, such as system controlled operable windows, or novel strategies to mix "non-pressurized" system ventilation with natural ventilation.

14. Provide fly fans or roll-up doors at entrances to the building, such as loading docks, to prevent unconditioned, uncontrolled outside air from entering the building.

15. Insulate pipes, ducts, plenums, and cold or hot equipment with the proper insulation types and thickness. Insulating pipes beyond the critical ratio will actually increase the quantity of heat transfer from the pipe.

16. Provide a vestibule for the main building entrance to prevent large amounts of unconditioned air from entering the building.

17. Provide overhangs for facades on buildings facing south, and adjustable customized fins for windows facing west and east to decrease the amount of solar heat entering the space.

18. External shading devices protect the building against excessive solar gains during the summer, but may have negative and positive effects on daylighting during the year. A daylight analysis should be used to show how positive effects can be maximized and glare can be minimized.

Building Internal Loads

1. Design outside air quantity based on real occupancy of the building. Do not over-estimate the number of people in the building. For a comfortable and hygienic indoor environment, a minimum ventilation rate is needed when the building is occupied, typically $0.3 * V$ [m³/hr], where V is the ventilated volume, in m³ for residential buildings, and 30m³/hr/person for non-residential buildings during occupancy period.

2. Use daylighting provisions to decrease the minimum required energy intensity of lighting. This decreases the size of the equipment and their associated ballast. However, when daylighting is utilized to reduce lighting electricity, the solar heat gain through glazing should be controlled, and in addition, glare and contrast must be controlled to provide a comfortable indoor environment. Daylights from skylights in North-facing zones are optimal.

For example, corridors in educational buildings are generally excellent spaces for daylighting. Overhangs should be positioned over the daylighting aperture and can possibly be sized with the light shelf to prevent direct sun from entering the space, especially during occupancy hours. Additionally, for office buildings, lower furniture in open plan office areas increases the efficiency of both the daylighting and the electric lighting system by reducing absorption and unwanted shadows. In office work areas, electric lights should be dimmed continuously rather than controlled manually in response to daylight to minimize employee distraction.

3. Use task lighting and occupancy sensors to decrease the local lighting load. The use of occupancy sensors with manual-on and automatic-off control in daylit spaces, such as classrooms, offices, mechanical rooms, and restrooms, saves lighting energy. Also, the use of local articulated task lights (desk lamps that can be adjusted in three planes) in daylit spaces increases occupants' satisfaction.

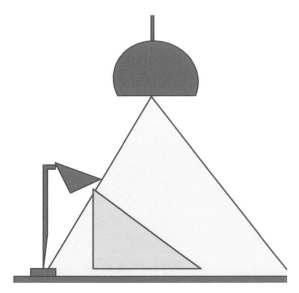

Figure 13   Use of Task Lighting to Decrease Local Lighting

4. Specify more efficient interior lighting. The widespread availability of compact fluorescent lamps and LED (light-emitting-diode) lighting options should reduce lighting electricity consumption and heat gains.

5. Try to apply methods of total light management, where external solar shading, internal shading, and electric lighting are controlled in a holistic manner.

6. Specify more energy efficient appliances to reduce the electricity requirements of plug loads and reduce heat gains from the usage of appliances, office equipment, and other devices plugged into electrical outlets.

[E.2] Energy Delivery Performance

SCOPE

COMMERCIAL | CORE + SHELL | RESIDENTIAL | EDUCATION | MOSQUES | HOTELS | LIGHT INDUSTRY | SPORTS

• Rated for SINGLE and GROUP RESIDENTIAL

DESCRIPTION

Establish delivered energy performance levels of the project in order to reduce environmental and economic impacts associated with excessive energy use.

GUIDELINES

EPC_{del} is a building performance indicator that is sensitive to the changes in a building's HVAC systems and other equipment.

HVAC Systems

When it is necessary to use on-site cooling generation equipment, such as chillers, control techniques should be considered; for example, variable volume flow, primary/secondary pumping, frame-and-plate heat exchangers use for water-side economizer, or use of chillers with de-superheating availability for use of wasted heat in chillers to warm-up any domestic water heating (in cases where this is required).

1. Using either an airside or waterside economizer usually provides options for large energy savings. However, based on the Qatar weather data (see Appendix II), the investment in a waterside economizer may not be a viable choice because local weather conditions are not suitable for the operation of a waterside economizer. On the other hand, an airside economizer requires the means to treat large amounts of outside air introduced into the building through the economizer cycle. This by itself can be a major source of unwanted air leakage during regular operation of the system. For these reasons, a detailed study should be conducted to justify the use of the economizer cycle in the project.

2. Use variable-speed-driven secondary pumps. This reduces pump energy by allowing each pump to operate at total system head. This also improves balancing of the system, and creates better part load performance.

3. Design and specify equipment based on operation near their maximum efficiency performance levels based on manufacturer data.

4. Use an energy recovery system to recover the heating or cooling from the exhaust air before discharging it to the outdoors.

5. Using variable air volume systems reduces the chance of over-cooling or over-heating a space when it is not at its peak load conditions. For particular applications, constant volume systems, such as fan coil units, can be used to provide better performance. Such systems are more efficient due to their smaller size, multiple units, and limited control requirements. Before selecting a system, a study must be conducted to specify which type of system is more appropriate for the specific application.

6. Avoid using temperature sensors that can be readjusted locally.

7. Use the maximum practical size of ductwork with a minimum amount of elbows and direction changes. This will reduce the air pressure loss inside the duct and therefore, minimize the required fan power for pushing air through the ducts.

8. Spaces that require continuous conditioning, such as electrical rooms or data centers, should be served separately from the rest of the building. This will eliminate the inefficient use of equipment in serving small, specific spaces when there is no demand for the rest of the building.

9. Use direct digital control systems to optimize start-up or shut-down of the systems.

10. Provide provisions to use wasted heat through boilers, if applicable, and water heaters' flue stacks for pre-heating water and air. However, this is typically not relevant in Qatar where no heating is required.

11. One of the most important factors in specifying HVAC systems for a building is designating the appropriate amount of outside air ventilation to the building, to provide a comfortable environment for occupants. Since this factor defines a major part of the load, particular attention should be paid to the design to provide the proper amount of outside air ventilation to the building. Excessive amounts of outside air will result in a high level of energy consumption, while a deficiency of outside air will make the building unhealthy and undesirable for the occupants.

12. Optimizing Chiller Design:

    12.1 Energy cost of a chiller plant depends on different variables such as:

        a. Full load and part load efficiency of elements (chillers, pumps, cooling towers, etc.) of the chiller plant.

        b. Design (variable flow or constant flow) and staging of the equipment.

        c. Control sequencing.

        d. Piping design.

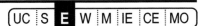
12.2 The ideal solution is to optimize all the elements at the same time, but it is not practical or possible to optimize all the elements simultaneously because of their interaction with each other.

12.3 Selecting the chilled water distribution flow arrangement: For a district cooling plant serving a group of large loads, such as campuses with multiple buildings, a Distributed Variable Speed Driven Secondary Pumps (DVSDSP) system (dedicated variable speed pump to each building) is the most efficient system. This is because it allows each secondary pump to be installed and to operate at system head required from the plant to the specific building, while in a conventional primary-secondary type chilled water system, a control valve is needed to throttle the pressure and distribute uniform pressure to the near and far loads. Using DVSDSP also improves performance in part load conditions.

12.4 Selecting control sequences for energy savings to minimize the energy use of the systems:

a. Stage Chillers & Pumps: Using variable speed chillers and pumping systems will ensure that the system efficiency does not drop until the load is down to about 20-25% (dependent on specific manufacturer's data). Strategies to help the efficiency of the system include the following:

i. For a single speed chiller/pumping system, it is more efficient to operate using only the minimal number of chillers necessary to satisfy the load. If the system operates with more chillers than needed to satisfy the load, the system will perform at lower efficiency.

ii. For a variable speed chiller/pumping system, it is more efficient to operate as many chillers as possible. Multiple chillers at lower loads perform better than a single chiller at near full load. Variable speed chillers make the staging process less complicated.

iii. For primary-secondary pumping systems, it is more efficient to start the chillers one by one to ensure that there is more primary chilled water than secondary chilled water in the loops.

iv. Use the plant Direct Digital Control (DDC) system to start/stop chillers, and use chiller controls (not the plant DDC system) to start/stop primary pumps to ensure additional energy savings.

b. Chilled Water Temperature Reset: Chillers are more efficient when the departing water temperature is higher. Increasing the chilled water temperature difference from the conventional 4-5 degree Celsius to as high as 10-11 degrees Celsius, decreases the required water capacity and therefore, decreases the size of the pumps and as a result, the motor power to run the pumps. A higher chilled water temperature difference causes higher airside pressure to drop and higher fan energy consumption, while simultaneously

decreasing the waterside pressure drop and as a result, the pump energy consumption drastically. A typical result of temperature changes on air and water pressure drop can be seen in Table 1:

Effect of Chilled Water Difference on Coil Pressure Drop			
Chilled Water Temperature Difference [°C]	5	8	10
Coil Water Pressure Drop, [kPa]	90	42	24
Coil Airside Pressure Drop, [kPa]	1.40	1.46	1.52

Table 1 Effect of Chilled Water Difference on Coil Pressure Drop

c. Thermal Storage: This system offers higher energy cost savings when the energy cost is different during the day and night.

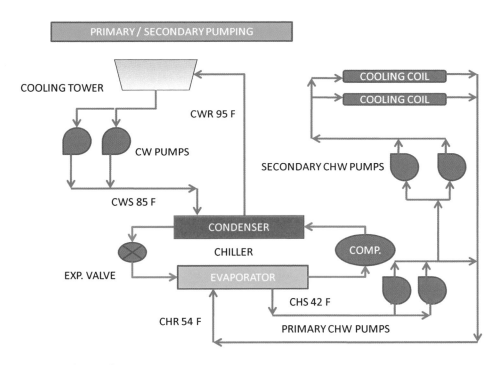

Figure 14   Primary/Secondary Pumping

12.5 Select valves for energy savings to minimize the energy use of a piping system:

a. Use ball valves or butterfly valves for insulation and balancing purposes, since both valves offer low pressure drop.

b. Do not install a strainer upstream from the coil to protect the control valve, because they usually generate flow problems. Strainers at the pump usually perform the required job and are very accessible.

13. Design the coil temperature difference about 2 to 3 degrees higher than the plant temperature difference (the water from the district cooling system) to account for coil heat transfer degradation.

14. Divide the floor plans into exterior and interior zones with HVAC systems serving each zone individually with its own temperature and humidity sensors, if applicable.

15. Provide motorized dampers for stairs and elevator shafts to reduce the chance of wasting conditioned air through these openings.

16. Provide motorized dampers for all intake and relief/exhaust louvers and vents to protect the conditioned air from leaving the building and prevent unconditioned outdoor air from coming into the building.

17. There is no need to provide freeze protecting devices for exterior pipes; but if so they should be shut off permanently.

18. Install cooling towers, chillers, and boilers and their associated pumps on the top floor of the building to decrease the required additional pump head and therefore, reduce the required motor power.

19. Use a multiple number of smaller chillers, instead of one or two large ones, to help the systems work most of the time around the full load situation and therefore, have greater efficiency.

20. Use condensate drain recovery from air handling unit coils to use for and reduce required cooling tower make up water or other water usages in the building.

21. Use fan powered terminal units to reuse the plenum heat in controlling the temperature of the space.

22. Use electrical finned tube radiation elements to protect large/tall lobby glasses.

23. In high-rise buildings, use multiple mechanical floors and multi-stage plate heat exchangers/pumps to reduce the required pump motor power.

24. Use flow measuring stations at outdoor air intake to the air handling units to control the quantity of the outdoor air.

25. Where possible, use one or more of the following design options to improve the efficiency of the system: primary-secondary pumping, ice storage, cogeneration, coil loop for exhaust air energy recovery, total energy recovery wheel, and direct or indirect evaporative cooling.

26. Where possible, use one or more of the following control strategies to improve the efficiency of the system: chilled and condenser water reset, fan cycling, demand limiting, duty cycling, and fan pressure optimization at part load operation.

27. Use condensing type boilers (if applicable) and domestic hot water heaters to increase the heat generation efficiency from the conventional 80% to up to 95-98%. In Qatar, this will only be appropriate in special circumstances where a heating system is required.

28. Electric centrifugal chillers have one of the highest efficiency ratings among other chillers. Use centrifugal chillers or other high efficiency chillers to increase the chiller efficiency to as high as 0.5 kW/ tons.

29. Where the chilled water piping volume relative to the chiller capacity is small, provide an inertia tank to protect the chiller operation against extreme on-off cycling.

30. Use innovative on-site energy generation methods, such as photovoltaic cells, to decrease the consumption of energy from public utility sources.

31. Provide provisions and connections for selling the additional on-site generated electricity to the city network, where building demand is satisfied but the city has additional demand.

32. Provide dedicated controlled exhaust systems for copy rooms to exhaust air from the room only when the copiers are functioning.

33. Provide adequate air intake space for outdoor air cooled equipment, such as cooling towers, to let them operate at the highest efficiency rating.

34. Prevent stratification of return air and outside air within the mixing box to improve the air handling unit efficiency.

35. Use chillers with de-super-heating ability or a heat exchanger to extract wasted heat from the chiller; and pre-heat the domestic hot water before entering hot water heaters, as illustrated in Figure 15 below.

36. Specify and enforce commissioning supervision for implementing all of the above mentioned improvements during the design phase.

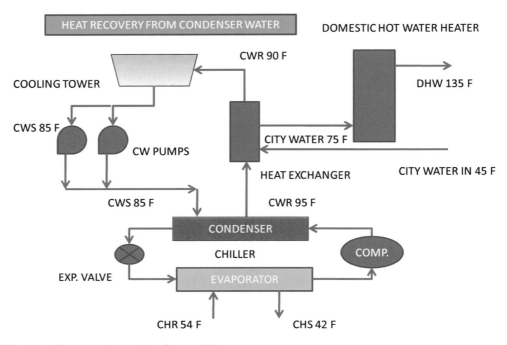

Figure 15   Heat Recovery From Condenser Water

A typical high-rise building was studied considering the local climate conditions in Qatar, with different HVAC systems as presented in Figure 16. The graph displays the relevant gains that result from different system choices. The most energy efficient system is the system with purchased district cooling. The DX packaged variable volume and air cooled chiller systems are about 10% less efficient. The other systems are 15-30% less efficient.

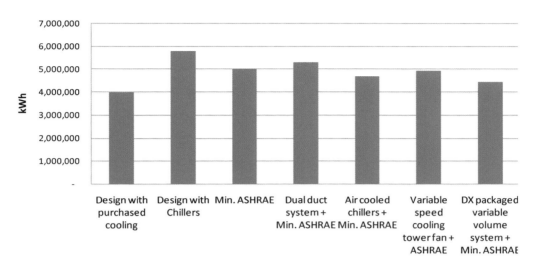

Figure 16   Different HVAC System Comparison

Renewable Strategies

Using photovoltaic cells to generate electricity can be an effective method for the generation of on-site electricity, and due to the length of day in Qatar, it is a viable opportunity that should be checked Additionally, the feasibility of other renewable energy sources, such as wind and ground source water loops for heat pumps, could be considered, but a thorough study of its true efficiencies and practicality need to precede their deployment.

1. Solar Energy and Photovoltaic (PV) Cells:

Solar energy systems can be used passively or actively to supply part of the building heating needs. In Qatar, there is no use for passive solar gain strategies. Rather, the "passive" design strategies should be focused on avoiding the entrance of solar gains as much as possible without negatively impacting access to daylight.

The best use of renewable systems is to generate electricity, for example, through the use of Photovoltaic (PV) panels or wind turbines. PV cells are sources of renewable energy that convert sun energy directly into electricity through the use of semiconductors.

Figure 17   Photovoltaic Panel

PV cells produce a direct current, which needs to be converted to an alternating current before it can be used in buildings. The common technique is to store currents from PV cells in batteries and change it to an alternating current through the use of an inverter. The major advantage of an alternating current created by the inverter is that the alternating current is compatible with the city utility grid. If the local PV system can generate additional electricity, the excess electricity can be transferred to the main utility grid and sold to the city electric provider, specifically when the city grid experiences high demand from customers during certain times of the day.

For early dimensioning purposes the following simple calculation process can be used:

> Using renewable on-site energy:
>
>> Roof size available: ~1,000m^2
>>
>> Area of each PV cell: 1.5 m^2
>>
>> Number of PV panels: 625
>>
>> Max. output of each PV panel: 200 watts
>>
>> Max. PV panels: 125 KW
>>
>> PV voltage: 270
>>
>> Estimated number of modules in series: 10
>>
>> Modules in parallel: 62
>>
>> Peak Amps of each module: 7.5
>>
>> Total Amps from PV panels: 465
>>
>> Average sun per day in Qatar: 8
>>
>> AMP-hour per day: 3720
>>
>> Watt-hour per day: 1004400
>>
>> Load Correction Factor: 1.1
>>
>> Power daily requirement: 910 kWh/day

In this case, the use of PV panels helps to decrease the average total daily energy consumption by 910 kWh in a building with 1,000m^2 of available roof area for installing PV cells. As the production of PV panels requires a large amount of energy, a life cycle analysis (LCA) is necessary to establish their net gain if a global perspective is needed.

2. Wind Power:

The incorporation of wind turbines is another option to generate power. The applicability largely depends on local wind conditions (the height of installation is a major factor) and the possibility to construct an ingoing concentrated airstream into the turbine or turbine array.

Turbine technologies are improving but installation costs are high. In the early stages the following calculation can be completed to check viability:

P_{total} (Watts)= Mass-flow rate (Kg/s) x $KE_{incoming}$ = 0.5 $(\rho A V i^3)/g_c$

P = Incoming wind density; Kg/m^3

V = Incoming wind velocity; m/s

A = cross-sectional area of stream; m^2

g_c = 1 Kg x m / (N x s^2)

The total power of a wind stream is directly proportional with its density, area, and its velocity to the third power.

The total power specified above cannot be converted to mechanical power completely and therefore, the maximum available power of a wind turbine can be shown as the following:

A wind turbine is only capable of converting 60% of the total power of wind to useful power at best:

P_{max} = 0.3 $(\rho A V i^3)/g_c$

η_{max} = P_{max} / P_{total} = 0.3 $(\rho A V i^3)/g_c$ / 0.5 $(\rho A V i^3)/g_c$ = 0.6

The above equation assumes ideal conditions along the entire wind turbine blades. Usually because of spillage and other effects, practical turbines achieve only 50-70% of the ideal efficiency. Therefore, the real efficiency η and Actual Power of a wind turbine are related to the total power via the following formula:

P_{actual} = η_{real} x P_{total} = 0.5 η_{real} $(\rho A V i^3)/g_c$

Where η_{real} is about 30-40% for the current turbines on the market.

Assume a 15 m/s wind in condition of (1 atmosphere pressure and 15°C) and a

10 m turbine diameter with ~40% real efficiency (η_{real}), the total actual produced power (P_{actual}) will be:

ρ (Air density) = P / RT =1.22 kg / m^3

P_{total} / A = 0.5 $(\rho V i^3)/g_c$ = 0.5 x (1.22 x 15^3)/1 = 2060 W/m^2

P_{actual} / A = η_{real} x P_{total} / A = 0.4 x 2060 = 824 W/m^3

P_{actual} = 824 x (3.14 x 102/4) = 64680 W = 64 KW

More thorough studies into the local situation need to be conducted to verify whether the use of wind turbines is a viable possibility for Qatar. The prospect of using local, building-attached or building-integrated turbines is at face value illogical because the same investment as part of a wind park elsewhere, such as in an optimal dedicated location, would produce a far larger amount of electricity for the same investment.

3. Ground Source Heat Pumps:

Ground source heat pumps use the unlimited capacity of the ground as the sink and source of heat rejection to provide energy efficient cooling. Figures 18 and 19 show cooling and heating cycles.

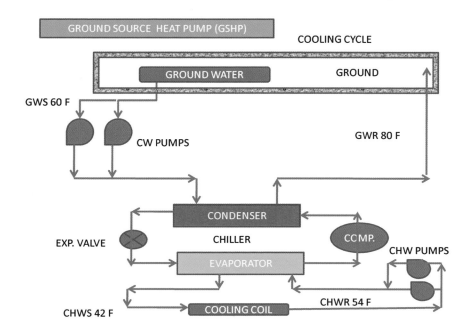

Figure 18   Ground Source Heat Pump for Cooling Cycle

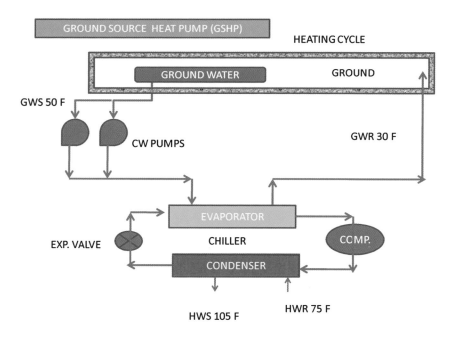

Figure 19   Ground Source Heat Pump for Heating Cycle

[E.3-E.5] Fossil Fuel Conservation, CO_2 Emissions, & NO_x, SO_x, & Particulate Matter

SCOPE

COMMERCIAL | CORE + SHELL | RESIDENTIAL | EDUCATION | MOSQUES | HOTELS | LIGHT INDUSTRY | SPORTS

- Rated for SINGLE and GROUP RESIDENTIAL

DESCRIPTION

Establish fossil energy conservation performance, CO_2 emission reduction performance, and NO_x, SO_x, and dust emission reduction performance of the building with its delivery systems and energy supply network.

GUIDELINES

EPC_p is a building performance indicator which is sensitive to changes in the method of building energy delivery. It depends on different types of energy delivery, using different types of energy supply networks such as electricity, gas, and district cooling. This factor is largely dependent on the city infrastructure, and in Qatar, the prevalent networks are the electricity from large, gas-fired power plants and chilled water delivery by district cooling plants.

Primary Energy Factor

After calculating the consumed energy at the building site, one should use the primary energy factor (PEF) to account for the efficiency of producing and delivering different types of energy to the site. In Qatar, natural gas is used as the main source of electricity generation, which is delivered to the doorstep of a building. The PEF would then be calculated as follows:

Assume a 15% loss for extracting gas from the ground, a 55% loss for converting gas to electricity, and a 10% loss for delivering the electricity to the building site. These numbers are completely based on how the infrastructure is designed in different countries, and for Qatar, a detailed study will have to be performed to determine usable numbers. For example, the PEF would then be calculated as (1-0.15)*(1-0.55)*(1-0.1) = 34.4%. Another example for the calculation for purchased chilled water depends on both chiller coefficient of performance (COP) and electricity resource utilization factor. Therefore, for a chiller with COP of 4 and 30% electrical resource utilization factor, the purchased chilled water resource utilization factor will be (4)*(0.30) = 120%.

These numbers only provide ballpark figures. Specific PEFs for energy carriers of electricity and thermal energy for Qatar have to be calculated using local data.

Emission Coefficient

CO_2, NO_x, and SO_x emission coefficients are used to estimate the impact of emissions from the energy delivered to a building. Emission coefficients are factors to measure emissions resulting from the primary resource inputs during fuel combustion at power plants. They vary depending on the type of resources used for electricity generation and the type of delivered energy as secondary energy from power plants. Emission coefficients represent the combination of conversion inefficiencies, and the transmission and distribution losses from the generation sources to the point of use. The conversion inefficiencies include the effects of pre-combustions, which are associated with extracting, processing, and delivering the primary resources to the point of conversions in the power plant or directly in the building. The EPC_{CO2}, $EPC_{NOx-SOx}$ values can be improved to have less emitting power supplies (which have less emission coefficient values) as an example introduced in the EPC_p improvement.

FURTHER RESOURCES

Publications:

1. *Energy Standard for Buildings Except Low-Rise Residential Buildings.* ASHRAE Standard 90.1-2004. Washington: American Society of Heating, Refrigerating and Air-Conditioning Engineers, 2004. Print.

2. Complainville, C. and J. O.-Martins (1994), "NOx/SOx Emissions and Carbon Abatement", *OECD Economics Department Working Papers, No. 151.* Organisation for Economic Cooperation and Development (OECD). OECD Publishing, 1994. Print.

3. Deru, M., & Torcellini, P. *Source Energy and Emission Factors for Energy Use in Buildings (No. NREL/TP-550-38617).* Golden: National Renewable Energy Laboratory, 2007. Print.

4. Dijk, D. v., & Spiekman, M. *CEN Standards for the EPBD - Calculation of Energy Needs for Heating and Cooling.* EPBD Buildings Platform, 2007. Print.

5. "Heating systems in buildings – Methods for calculation of system energy requirements and system efficiencies, Part 3-1: Domestic hot water systems, characterization of needs (tapping requirement)." EN 15316-3-1. Brussels: European Committee for Standardization (CEN). Print.

6. "Heating systems in buildings – Methods for calculation of system energy requirements and system efficiencies, Part 4-6: Heating generation systems, photovoltaic systems." EN 15316-4-6 Brussels: European Committee for Standardization (CEN). Print.

7. "Energy performance of buildings - Overall energy use and definition of energy." EN 15603. Brussels: European Committee for Standardization (CEN). Print.

8. "Energy performance of buildings - Calculation of energy use for space heating and heating." EN-ISO 13790. Brussels: European Committee for Standardization (CEN) and International Organization for Standardization (ISO). Print.

9. *District Heating - Heating More with Less.* Brussels: Euroheat & Power, 2005. Print.

10. *Energy Performance of Non-Residential Buildings. Determination Method.* NEN 2916:1998. Delft: Nederlands Normalisatie-instituut, 1998. Print.

11. "Energy performance of buildings - Energy requirements for lighting." PrEN 15193. Brussels: European Committee for Standardization (CEN). Print.

12. "Energy performance of buildings - Methods for expressing energy performance and for energy certification of buildings." PrEN 15217. Brussels: European Committee for Standardization (CEN). Print.

13. "Energy performance of buildings Impact of building automation, controls and building management." PrEN 15232. Brussels: European Committee for Standardization (CEN). Print.

14. "Ventilation for buildings - Calculation methods for energy losses due to ventilation and infiltration in commercial buildings." PrEN 15241. Brussels: European Committee for Standardization (CEN). Print.

15. "Ventilation for buildings - Calculation methods for the determination of air flow rates in buildings including infiltration." PrEN 15242. Brussels: European Committee for Standardization (CEN). Print.

16. "Indoor Environmental input parameters for design and assessment of energy performance of buildings addressing indoor air quality thermal environment, lighting and acoustics." PrEN 15251. Brussels: European Committee for Standardization (CEN). Print.

17. "Explanation of the general relationship between various European standards and the Energy Performance of Buildings Directive (EPBD) - Umbrella document." TR 15615. Brussels: European Committee for Standardization (CEN). Print.

18. *Buildings and Climate Change: Status, Challenges and Opportunities*. Paris: United Nations Environment Program, 2007. Print.

WATER [W]

With minimal rainfall, a hot and dry climate, and the depletion of groundwater at rapid rates, water has become a precious and limited resource in the region. The project should aim to reduce the demand on fresh water consumption as well as recycle, treat, and reuse water where feasible. Promote the efficient use of water in the project to mitigate environmental impacts associated with water scarcity and depletion. Strategies for consideration include the following: using efficient technologies, equipment, and fixtures; collecting, treating, and reusing greywater, rainwater, and sewage; minimizing water pollution; and properly managing and maintaining water-use systems. Incorporate as many strategies into the project as possible to reduce fresh water demand.

Criteria in this category include:

W.1   Water Consumption

[W.1] Water Consumption

SCOPE

COMMERCIAL | CORE + SHELL | RESIDENTIAL | EDUCATION | MOSQUES | HOTELS | LIGHT INDUSTRY | SPORTS

- Rated for SINGLE and GROUP RESIDENTIAL

DESCRIPTION

Minimize water consumption in order to reduce the burden on municipal supply and treatment systems.

GUIDELINES

Reduce overall water consumption during building operations using the following strategies: specification of water efficient equipment and fixtures, reuse of rainwater and greywater for non-potable applications, use of on-site sewage treatment systems, installation of water use sub-metering facilities, educating staff on water conservation, use of efficient landscape irrigation techniques, and planting of native, low water consuming vegetation. Consider the various options and methodologies to maximize water efficiency based on project specific goals and requirements, as well as site, location, applicable codes, building typology, and hours of operation.

Water Efficient Equipment and Fixtures

Consider specifying water efficient equipment such as low flush toilets, vacuum toilet flush systems, dual flush toilets, flow-controllers and regulators, water-saving valves and fixtures on faucets and showerheads, low flush urinals, low-water dishwashers, and occupant sensors. Specify automatic shut-offs, electronic sensors, and low-flow or lever taps on faucets. Dry fixtures such as composting toilets and waterless urinals can also be selected to reduce water demand. Low-flow plumbing fixtures and appliances should be specified instead of conventional fixtures and appliances to reduce occupant water consumption. Leak detection systems should be installed to quickly and efficiently identify and locate water leakage points.

If the project has dining facilities, restaurants should provide ice water only on request in order to reduce overall ice consumption. In addition, specify air-chilled ice machines instead of water-chilled models. Installing high-efficiency dishwashers with rinse water recycling systems may contribute to additional water savings. Flight or conveyor dishwashers are most efficient, followed by undercounter dishwaters. Manual dishwashing should be avoided because it consumes more water than automatic dishwashers.

If the project has laundry facilities, the building should adopt laundry policies to use water more efficiently.

Maximize water heating efficiency by separating the hot water system for laundry from other building water systems, if possible. Operate washers only at full capacity, and use the minimum amount of required detergent. To reduce the overall use of hot water and the need for harmful chemicals such as chlorine bleach, install ozone washing systems, or install high-efficiency washing machines with rinse water recycling systems.

Rainwater and Greywater Reuse

Design the project to collect, store, and redistribute rainwater and greywater on the project site to reduce the building's potable water consumption. The treatment of rainwater and greywater is dependent upon the quality of the water; they do not necessarily have to be treated prior to redistribution. Greywater includes water discharge from building operations such as condensate from cooling systems, laundry, bathroom sinks, water discharged from cooling towers, and water fountains. Greywater generated from kitchen and catering facilities should be stored and treated separately from other sources because oils and fats are difficult to remove. Dual plumbing lines can be used to separate greywater from sewage water and should be installed during initial construction.

Collect water generated by the operation of air conditioners, dehumidifiers, and refrigeration units. Condensation does not necessitate a stringent treatment process but can contain contaminants and bacteria, and should only be reused for non-potable applications in the building and on the site.

Rainwater can be harvested and collected using devices such as cisterns or underwater tanks. Rainwater collected from roofs and other impervious surfaces on the site can be filtered using various methods, including screens and paper filters. Water collected from paved surfaces, such as roads and parking lots, may require oil separators and further treatment to eliminate oils, fuels, and other harmful substances. Reuse on-site greywater and rainwater for non-potable applications including toilet/urinal flushing, irrigation, custodial/janitorial uses, fire protection, mechanical systems, and car washing.

Sewage Treatment

Treating sewage on-site creates recycled water and sludge, which is the byproduct of waste. Recycled water can be used for both potable and non-potable uses, depending on the type of treatment utilized. Recycling water produced from sewage can minimize water demand and consumption from building operations and reduce the burden on public treatment facilities. Sludge may be taken to appropriate disposal facilities or biologically digested on-site to produce methane. Use separate plumbing lines for sewage to isolate water from other wastewater systems. Consider innovative methods, such as low-pressure dosing systems with septic tanks and use techniques such as solid separators, sand filters, aerobic tank treatment, and aquaculture systems, to treat sewage. Sand filters are well-established sewage treatment systems and are extremely cost-effective. Aquaculture systems are biological techniques that utilize plants and fish to ingest pollutants in wastewater, in turn, producing purified water, food, and fertilizer. Such systems require more maintenance to ensure proper functioning but may be an advantageous method for treating sewage. The various treatment systems should be researched and analyzed for the proposed project, to meet its specific goals and requirements.

Water Meters

Water consumption can be reduced if its use is regularly monitored and managed during building operations. Install water meters on the main water supply to each building in the proposed project. Water meters should be easily accessible and convenient for facilities operators. Sub-metering of water uses is also recommended for high water-usage operations, such as boilers and cooling towers, to ensure proper management and minimize water consumption. Monitor irrigation systems to control over-watering and to detect the build-up of nutrients such as nitrogen, calcium, potassium, and sodium. Consider connecting the water meter to the building monitoring system using a pulsed output to ensure detection of inefficiencies in water use and consumption.

Efficient Irrigation

Reduce the need for irrigation through the specification of native or adapted plants that require nominal or no irrigation after initial establishment. Shade the site where possible to minimize water loss due to evaporation. Where irrigation systems are necessary, they should be regularly maintained and managed to ensure effective operation. Minimize the use of potable water for irrigation using harvested/recycled rainwater and greywater where feasible. Group plants with similar water needs for the most efficient use of water, and develop and implement appropriate watering schedules. Consider efficient, low-water irrigation systems such as drip feed subsurface systems, and utilize weather-based irrigation strategies such as rain shut-offs, moisture sensors, and evapotranspiration/smart irrigation controllers. Irrigation techniques for greywater may include drip feed subsurface systems, traditional evapotranspiration systems, and shallow trench systems that allow for subsurface irrigation of plant roots. If feasible, merge the greywater reuse system with the irrigation system in order to reduce the need to treat greywater on-site as well as to reduce the need for potable water for irrigation use. Specify a subsurface irrigation system when using greywater to avoid possible risks to human health. Avoid the activation of irrigation systems during the day, and utilize mulch and/or gravel to prevent water evaporation from the soil.

Water Features

Limit water features in and around the building or development in order to conserve water, and use recycled water for recirculation. Design water features with trickling or cascading fountains as they lose less water to evaporation than those spraying water into the air. Avoid placing water features outdoors due to the loss of water from evaporation. Indoor and outdoor water features should be operated on separate systems to prevent additional water loss. Also consider using salt water in water features to reduce the need for desalinated water, but ensure that water feature materials such as tile and stone can withstand the corrosive properties of salt.

Staff Education & Maintenance Plans

Dishwashing and kitchen water consumption can be limited by posting educational information, holding seminars, and instructing staff members to consciously reduce overall water consumption during building operation. Educate kitchen staff regarding correct methods for thawing frozen food and rinsing foods. Sub-meter kitchen water use and back charge costs to the kitchens to help raise awareness about wasteful practices among kitchen managers.

Develop strategic maintenance policies and procedures to ensure that water conservation is a focus in all building operations including cleaning, maintenance, and food service. Custodial staff should reduce water usage and the number of times toilets are flushed during room cleaning. Additionally, power washing the building exterior should be done on an as-needed basis separate from the regular maintenance plan. Power washing should not be used for general cleaning purposes, such as spraying down loading docks, when sweeping or spot cleaning will suffice.

EDUCATION

Water saving strategies, including reuse systems, can be used to instruct students about the importance of water in the region. Incorporate water saving strategies into the design that allows students to participate in conservation.

HOTELS

Linen Reuse Program

In order to reduce overall water consumption, introduce a linen reuse program that will encourage hotel guests to reuse towels, sheets, pillow cases, restaurant linens, and other linens. Increasing the lifespan of linens, in addition to reducing water and chemical detergent consumption, can provide significant environmental and economic benefits for the hotel. Ensure that guests are aware of linen reuse through signage or published information in guestrooms. Allowing guests to choose to conserve water by reusing towels and not having sheets changed daily can contribute to a positive experience in the hotel stay as well as reduce water consumption.

Water Efficient Equipment and Fixtures

Strategically reduce the total number of ice machines to reflect average demand of hotel occupants. Restaurants and room service should provide ice water only on request in order to reduce overall ice consumption. In addition, specify air-chilled ice machines instead of water-chilled models.

HOTELS | SPORTS

Recreational and Competition Venue Components

Recreational and competition venue components consist of areas of the sports facility or hotel complex that cannot reduce water consumption without affecting the requirements of the sport or recreational activity. Examples of such venues include, but are not limited to, indoor pools, ice rinks, or golf courses. Install sub-meters to monitor water usage for all pool, medical, and spa facilities. Limit pool and spa drainage by maintaining the chemical balance of pool water. Install chemical regulators to avoid human error when balancing chemicals. Pools may only require annual drainage, whereas spas require more frequent drainage, which can be further reduced through proper monitoring of chemical composition. If applicable, install flow regulators within pools, hot tubs, or steam rooms to reduce water consumption from those devices. Post signage surrounding pool deck and hot tub rooms requesting users to limit their pool and spa facilities to a designated time. Due to the intense energy required for desalination, consider using salt water pools. Saltwater pools reduce the need to purchase, store, add, and handle chlorine by converting the salt to chlorine through a chlorinator device. Because salt can be corrosive over time to metals like stainless steel or unsealed stone and cement, pool areas should be properly maintained and constructed of corrosion resistant materials. In addition, avoid landscaping in splash back areas around salt water pools as the saline may damage plants.

Landscaped areas used for recreational activities and competition, such as golf courses and soccer fields, should adopt landscape management practices that conserve water, and course design should reflect water conservation policies such as limiting lawn areas and eliminating water hazards. For irrigation, use non-potable water sources such as harvested water or greywater recycled from hotel and/or facilities use. Further reduce the need for irrigation by installing water retention systems underneath the soil, and cover putting greens with tarps during evening hours to trap condensation.

LIGHT INDUSTRY

Water Efficient Industrial Equipment and Processes

Specify equipment used in the industrial process to use water resources efficiently. Equipment should use as little water as possible and reduce excess consumption. Equipment should be designed and installed to reduce the potential for leaks and spillage. Additionally, equipment should recycle and reuse water when possible.

SPORTS

Functional components and competition components can be reviewed for water efficiency using all the techniques listed in the overall water consumption reduction methods as well as project specific items such as swimming pools, ice rinks, and golf courses.

Water Efficient Equipment and Fixtures

For legacy use, install shut off valves and plan to evacuate water from pipes in unused areas during smaller events which will decrease water loss and increase efficiencies.

Efficient Irrigation

Fresh water consumption should be avoided by using recycled water for irrigation in competition areas where changes in vegetation and shading are detrimental to the functional requirements of the competition venue.

Facility Functional Components

The functional components of a sports facilities which may require special attention in regards to water consumption are the laundry facilities, fitness suites, restaurant kitchen areas, common areas, and animal facilities.

Fitness suites can further reduce water consumption by installing sub-meters to monitor water usage for all medical hot tubs and steam rooms. Medical hot tubs require frequent drainage, which can be reduced by installing multiple person tubs and by properly monitoring the chemical composition. If applicable, install flow regulators in hot tubs or steam rooms, as well as request users to limit time spent consuming fresh water.

FURTHER RESOURCES

Websites:

1. *American Rainwater Catchment Systems Association.* American Rainwater Catchment Systems Association. Web. 3 June 2010. <http://www.arcsa.org/index.html>.

2. United Kingdom. Department for Environment, Food and Rural Affairs. *Department for Environment, Food and Rural Affairs.* Web. 30 June 2010. <http://www.defra.gov.uk/environment/quality/water/conserve/index.htm>.

3. *International Rainwater Catchment Systems Association.* International Rainwater Catchment Systems Association. Web. 3 June 2010. <http://www.eng.warwick.ac.uk/ircsa/>.

4. *The Irrigation Association.* Irrigation Association, 2010. Web. 3 June 2010. <http://www.irrigation.org>.

5. *The UK Rainwater Harvesting Association.* The UK Rainwater Harvesting Association. Web. 3 June 2010. <http://www.ukrha.org/>.

6. *Rainharvesting Systems.* Rainharvesting Systems. Web. 3 June 2010. <http://www.rainharvesting.co.uk/>.

7. *Water-Efficient Landscaping.* University of Missouri Extension, 1993. Web. 04 June 2010. <http://muextension.missouri.edu/xplor/agguides/hort/g06912.htm>.

8. *Savewater.* Savewater, 2005. Web. 3 June 2010. <http://www.savewater.com.au/>.

9. Hong Kong. The Government of the Hong Kong Special Administrative Region of the People's Republic of China. *Water Supplies Department .* Web. 3 June 2010. <http://www.wsd.gov.hk>.

10. Australia. National Program for Sustainable Irrigation. *National Program for Sustainable Irrigation.* Web. 3 June 2010. <http://www.npsi.gov.au/>.

Publications:

1. United Kingdom. Environment Agency. *Conserving Water in Buildings – A Practical Guide.* United Kingdom: Environment Agency, 2007. Print.

2. United States. Environmental Protection Agency. *On-Site Wastewater Treatment Systems Manual.* Washington: EPA , 2002. Print.

3. United States. Environmental Protection Agency. *Reclaimed Water Systems: Information about Installing, Modifying or Maintaining Water-Efficient Landscaping: Preventing Pollution & Using Resources Wisely.* Washington: EPA, 2002. Print.

4. *Rainwater and Greywater Use in Buildings, Best Practice Guidance.* Construction Industry Research and Information Association, 2001. Print.

5. *Reclaimed Water Systems: Information about Installing, Modifying or Maintaining Reclaimed Water Systems 9-02-04.* Water Regulations Advisory Scheme, 1999. Print.

6. *Conservation of Water, 9-02-03.* Water Regulations Advisory Scheme, 2005. Print.

7. United Kingdom. *Water Supply (Water Fittings) Regulations.* Department for Environment, Food and Rural Affairs, 1999. Print.

8. Brown, Reginald and Anu Palmer. *Water Reclamation Guidance: Design and Construction of Systems Using Grey Water, TN 6/2002.* United Kingdom: Building Services Research and Information Association, 2002. Print.

9. Brown, Reginald and Anu Palmer. *Water Reclamation Standard: Laboratory Testing of Systems Using Grey Water, TN 7/2002.* United Kingdom: Building Services Research and Information Association, 2002. Print.

10. Pidou, Marc, et al. "Greywater Recycling: A Review of Treatment Options and Applications," *ICE Proceedings: Engineering Sustainability.* 160. United Kingdom: Institution of Civil Engineers, 2007. 119-131. Print.

11. Smith, Stephen. *Landscape Irrigation: Design and Management.* New York: John Wiley and Sons, 1996. Print.

12. Villacampa Esteve, Y., C.A Brebbia, and D. Prats Rico, Eds. "Sustainable Irrigation Management, Technologies and Policies." *WIT Transactions on Ecology and the Environment.* United Kingdom: WIT Press, 2008. Print.

13. United States. City of Seattle. Seattle Public Utilities: Resource Conservation Section. *Hotel Water Conservation: A Seattle Demonstration.* Seattle: Seattle Public Utilities, 2002. Print.

MATERIALS [M]

Ensure that the design of the building minimizes the unnecessary use of materials in order to mitigate environmental impacts associated with material use. Consider the following factors to mitigate environmental impact due to material use: using regional and local materials to reduce transportation needs; using responsibly sourced materials; using materials with high recycled contents and recycling materials on- and off-site; reusing materials and structure from on- and off-site sources; designing for ease of disassembly and reuse/recycling; and selecting materials with low life cycle environmental impact. Select the appropriate materials for the project based on specific requirements and goals as well as the overall impact to the environment.

Criteria in this category include:

M.1	Regional Materials
M.2	Responsible Sourcing of Materials
M.3	Recycled Materials
M.4	Materials Reuse
M.5	Structure Reuse
M.6	Design for Disassembly
M.7	Life Cycle Assessment (LCA)

[M.1] Regional Materials

SCOPE

COMMERCIAL | CORE + SHELL | RESIDENTIAL | EDUCATION | MOSQUES | HOTELS | LIGHT INDUSTRY | SPORTS

- Rated for SINGLE and GROUP RESIDENTIAL

DESCRIPTION

Encourage the use of regionally manufactured and assembled building elements and materials in order to reduce the carbon footprint of the materials.

GUIDELINES

Use locally manufactured and assembled materials and products to support the local economy and reduce impacts on the environment resulting from transportation. Where available, high weight materials such as aggregate, concrete, masonry, sand, and steel should be locally sourced. Since heavier materials require more energy to transport, they have greater impact on the environment if sourced from outside the region. To be considered locally sourced, materials and products must be assembled as a finished product within the region.

Projects should inquire with local or regional agencies for a list of manufacturers within the region and the products or services they provide. Determine which of these products or materials can be used and purchased, and then run budget calculations to set a target for the specific project. These considerations should take place early in the design process to assess which locally sourced products or materials will be most appropriate and feasible in terms of the project design and budget.

Consider creating a Materials Logistic Plan to take into consideration sourcing distances of manufacturers, security and storage of materials on-site, as well as use, recycling, or disposal of waste. For more information on materials logistic planning and the benefits of regional materials, use the tools and guidance section provided by Waste and Resources Action Programme (WRAP) in the United Kingdom.

HOTELS

Because hotels install large quantities of permanent interior fixtures and finishes, these items should also be sourced from within the region in order to support the local economy and further reduce environmental impacts.

FURTHER RESOURCES

Websites:

1. *Construction; Tools and Guidance.* Waste and Resources Action Programme, 2010. Web. 17 August 2010. <http://www.wrap.org.uk/construction/tools_and_guidance/index.html>.

[M.2] Responsible Sourcing of Materials

SCOPE

COMMERCIAL | CORE + SHELL | RESIDENTIAL | EDUCATION | MOSQUES | HOTELS | LIGHT INDUSTRY | SPORTS

- Rated for SINGLE and GROUP RESIDENTIAL

DESCRIPTION

Encourage the use of responsibly sourced materials for primary building elements in order to minimize the depletion of nonrenewable materials.

GUIDELINES

Use responsibly sourced materials for key building elements to conserve natural resources and to reduce the environmental impact of extracting and processing nonrenewable materials. Materials that have achieved certification by their supplier or manufacturer as being responsibly sourced are qualified for consideration in this criterion.

The project's key building elements, including the structural frame, ground floor, upper floors, roof, external/internal walls, foundation/substructure and staircases, should be comprised of responsibly sourced materials. Materials to be responsibly sourced include the following:

Brick, clay tiles, and other ceramics; resin-based composites and materials including GRP and polymeric render; concrete; mortars; cementitious renders; glass; plastics and rubbers (including EPDM, TPO, PVC, and VET roofing membranes); metals; building stone; timber and wood panel (MDF, chipboard, and particleboard); plasterboard and plaster; bituminous materials (roofing membranes and asphalt; other mineral-based materials (including fiber cement and calcium silicate); thermal insulation materials; and materials and products with recycled content.

Consider specifying rapidly renewable materials in the project such as bamboo, wool, cotton insulation, soy-based solvents and insulation, agrifiber, linoleum, wheatboard, strawboard, and cork. Rapidly renewable materials have harvest cycles of less than ten years and are less likely to become depleted from overharvesting. However, it is prudent to research the durability and longevity of such materials prior to selecting them for the project.

Identify the responsibly sourced materials and suppliers that are available to the project. Determine which of these materials can be used and purchased, and then run budget calculations to set a target for the specific project. These considerations should take place early in the design process to assess which responsibly sourced materials will be most appropriate and feasible in terms of the project design and budget.

To ensure the selection of socially and environmentally conscious materials that employ responsible practices throughout the supply chain, all imported materials must comply with the following third-party standards, where applicable:

Social Responsibility Policies

Materials should follow the standards established by ISO 26000, which defines core issues, principles and practices relating to integrating, implementing and promoting socially responsible behavior in organizational practices. Choose suppliers by their compliance with proper working conditions, environmental practices, safety standards, and human rights policies. Before sourcing goods from low-income countries, conduct a risk analysis to identify environmental and labor issues and to assess cost-analysis benefits. Take initial steps to research the environmental and social legislation in countries from which goods are imported to assess risks. Ensure compliance of suppliers with national laws and regulations, including labor and environmental laws, through contracts or industry-wide supplier codes of conduct. Encourage suppliers to develop their own responsible practices by emphasizing the benefits of responsible business practices on quality, productivity, contract renewals, and lowering employee turnover. Also, give preference to companies that provide training programs for management and employees that cover supervisory skills, environmental management, and health and safety awareness.

Quality Assurance & Supply Chain Management Systems

Materials should originate from sources with a documented quality management system based on the standards of ISO 9001, and materials purchased should follow the guidelines established by Sub-clause 7.4 of ISO 9001. If accredited systems already exist for specific materials, those standards should be followed. Choose suppliers based on their ability to deliver goods through analysis of supply chain issues, product quality and safety, continuity of supply and speed of delivery, and intellectual property protection. Avoid undermining the capacity of suppliers to respect social and environmental standards by placing rush orders, last-minute changes, or placing orders that surpass suppliers' capabilities, which contributes to inefficiencies such as excessive overtime work and other compliance violations. Ensure that a company's direct second and third tier suppliers are observing responsible social and environmental practices further down the supply chain.

Environmental Management Systems

Materials used in the project should originate from sources with a documented Environmental Management System (EMS). The EMS should cover the entire supply chain including acquiring and extracting raw materials, material manufacturing and associated processes, and related energy consumption. The EMS should adhere to the principles covered in ISO 14001. Ensure that products originate from companies that make the environment an organizational priority and integrate environmental management practices in order to identify problems and prevent recurrence. Organize on-site visits, conduct worker interviews, and hire independent monitors to conduct assessments of suppliers' facilities and practices for compliance.

Timber

The overharvesting of forests has led to the extinction of many tree species and the depletion of wood as a natural resource. All timber and wood products should originate from sustainably managed forests. All timber must be supplied by companies that hold Forest Stewardship Council (FSC) Chain of Custody Certification. Products should originate from forest management companies that comply with national legislation, demonstrate long–term land tenure and use rights, recognize the rights of indigenous people, maintain the ecology and biodiversity of the forest, enhance economic viability, and conduct adequate management, planning, and monitoring of operations.

FURTHER RESOURCES

Websites:

1. *Green Building Materials.* California Department of Resources Recycling and Recovery. Web. 14 June 2010 <http://www.ciwmb.ca.gov/GreenBuilding/materials/>.

2. *The Global Conservation Organization.* World Wildlife Federation. Web. 30 June 2010. <http://www.panda.org>.

3. *Convention on International Trade in Endangered Species of Wild Flora and Fauna.* Convention on International Trade in Endangered Species. Web. 30 June 2010. <http://www.cites.org/>.

4. *ICC guidance on supply chain responsibility.* International Chamber of Commerce. Web. 30 January 2011. <http://www.iccwbo.org/uploadedFiles/ICC/policy/business_in_society/Statements/141-75%20int%20rev6%20FINAL.pdf>.

5. *ILO Declaration on Fundamental Principles and Rights at Work and its Follow-up.* International Labour Organization. Web. 30 January 2011. < http://www.ilo.org/declaration/thedeclaration/textdeclaration/lang--en/index.htm>.

6. *Tripartite declaration of principles concerning multinational enterprises and social policy (MNE Declaration) - 4th Edition.* International Labour Organization. Web. 30 January 2011. <http://www.ilo.org/wcmsp5/groups/public/---ed_emp/---emp_ent/---multi/documents/publication/wcms_094386.pdff>.

7. *The Universal Declaration of Human Rights.* United Nations. Web. 30 January 2011. <http://www.un.org/en/documents/udhr/index.shtml>.

Publications:

1. *Environmental Impact of Building and Construction Materials, SP116.* United Kingdom: Construction Industry Research and Information Association, 1995. Print.

2. *Guidance on social responsibility.* ISO 26000:2010. International Organization for Standardization, 2010. Web. 30 January 2011. <http://www.iso.org/iso/iso_catalogue.htm>.

3. *Quality management systems -- Requirements.* ISO 9001:2008. International Organization for Standardization, 2008. Web. 30 January 2011. <http://www.iso.org/iso/iso_catalogue.htm>.

4. *Environmental management systems -- Requirements with guidance for use.* ISO 14001:2004. International Organization for Standardization, 2004. Web. 30 January 2011. <http://www.iso.org/iso/iso_catalogue.htm>.

5. *FSC Standard for Chain of Custody Certification.* FSC-STD-40-004 (Version 2-0) EN. Forest Stewardship Council, 2004. Web. 30 January 2011. <http://www.fsc.org/fsccertification.html>.

[M.3] Recycled Materials

SCOPE

COMMERCIAL | CORE + SHELL | RESIDENTIAL | EDUCATION | MOSQUES | HOTELS | LIGHT INDUSTRY | SPORTS

• Rated for SINGLE and GROUP RESIDENTIAL

DESCRIPTION

Encourage the use of building elements and materials made from recycled content in order to reduce the need for virgin materials.

GUIDELINES

Use materials with recycled content to reduce the environmental impact of extracting and processing nonrenewable and virgin materials. Materials, products, and furnishings made with recycled content can contain post- or pre-consumer content. Post-consumer waste is content that is recycled after the product has been used by a consumer. It may include construction or demolition debris such as recycled aggregate, steel and aluminum building elements, materials sorted for recycling purposes such as aluminum cans and glass bottles, and landscaping waste such as branches and leaves. Pre-consumer waste is raw material that has never been used by a consumer such as wood chips and sawdust. Pre-consumer materials are oftentimes by-products of manufacturing processes that can be recycled and reused.

Identify the recycled materials that are available and their respective suppliers. Ensure that the recycled materials selected for the project are of a high quality, have no detrimental environmental impacts, and will not hinder construction in any way. Determine which of these materials can be used and purchased, and then run budget calculations to set a target for the specific project. These considerations should take place early in the design process to assess which recycled materials will be most appropriate and feasible in terms of the project design and budget. Verify and document the post- and pre-consumer content of each selected recycled material through manufacturer or supplier specifications.

EDUCATION

Vinyl Composite Tiles (VCT) are often used for floors in education buildings. An alternative is tile made from recycled rubber. When using tile from recycled rubber, check to make sure it is not a strong volatile organic compound (VOC) emitter. Specify the minimum amount of adhesive necessary to have proper performance and insure proper ventilation during installation. Recycled rubber is also a good material for running tracks and playgrounds.

Insulation, acoustic wall panels, and ceiling tiles made from recycled materials are widely available. Gypsum board should have 100% recycled content paper facing. The gypsum board core should be a minimum of 10% recycled content. Unpainted gypsum board and acoustic ceiling tiles easily absorb VOCs. When possible, construction and installation should be sequenced to avoid using VOC-containing materials after acoustic panels and tiles are installed.

HOTELS

Because hotels install large quantities of permanent interior fixtures and finishes in addition to building materials, these items should contain recycled content when possible.

FURTHER RESOURCES

Websites:

1. *Wastes - Resource Conservation - Comprehensive Procurement Guidelines.* United States Environmental Protection Agency, 2009. Web. 30 June 2010. <http://www.epa.gov/epawaste/conserve/tools/cpg/index.htm>.

[M.4] Materials Reuse

SCOPE

COMMERCIAL | CORE + SHELL | RESIDENTIAL | EDUCATION | MOSQUES | HOTELS | LIGHT INDUSTRY | SPORTS

- Rated for SINGLE and GROUP RESIDENTIAL

DESCRIPTION

Encourage the reuse of building elements and materials in order to reduce the need for virgin materials.

GUIDELINES

Salvage, refurbish, or reuse on- or off-site building materials and products in order to divert material from the waste stream and reduce the environmental impacts associated with producing new materials and products. Salvaged materials are those materials taken from existing buildings and reused in new buildings and developments. Salvaged materials include structural beams and posts, flooring, paneling, windows, doors and frames, cabinetry, furniture, and masonry. They can be purchased from suppliers or recovered and relocated directly from an existing building.

Identify the salvaged building materials and products that are available and their respective suppliers or locations. Determine which of these materials can be used and purchased, and then run budget calculations to set a target for the specific project. These considerations should take place early in the design process to assess which salvaged materials will be most appropriate and feasible in terms of the project design and budget.

FURTHER RESOURCES

Publications:

1. *FSC-STD-40-007 V1-0 EN: Sourcing reclaimed materials for Use in FSC Product Groups or FSC-certified Projects.* Forest Stewardship Council, 2007. Web. 30 January 2011. <http://www.fsc.org/fileadmin/web-data/public/document_center/international_FSC_policies/standards/FSC_STD_40_007_V1_0_EN_Sourcing_reclaimed_materials.pdf>.

2. *The Reclaimed and Recycled Construction Materials Handbook, C513.* United Kingdom: Construction Industry Research and Information Association, 1999. Print.

3. Ohio. Environmental Protection Agency. *Pollution Prevention by Building Green, No 86.* Columbus: Office of Compliance Assistance and Pollution Prevention, 2004. Print.

[M.5] Structure Reuse

SCOPE

COMMERCIAL | CORE + SHELL | EDUCATION | HOTELS | LIGHT INDUSTRY

DESCRIPTION

Encourage the reuse of structural elements in order to reduce the need for virgin materials.

GUIDELINES

All projects should reuse structural elements originating from on- or off-site sources to reduce the impacts associated with the extraction and processing of virgin and nonrenewable resources. Reusing existing structural elements also decreases the amount of construction waste produced and reduces environmental impacts associated with transportation.

In cases where there are existing structural elements on-site, determine whether the existing elements are in usable condition and remove any elements that may be harmful to future occupants. Possible structural elements to be reused include the foundation and slab, basement and retaining walls, exterior walls, structural system, floors, and interior bearing walls. Reuse of the existing structural elements should not necessitate significant strengthening or alterations to make them structurally viable.

Projects that reuse any structural elements should generate a list and area of all major structural elements required for the project noting the reused elements, as well as all relevant drawings that document the extent of the structural elements to be reused. Determine whether the elements are reusable and/or appropriate for the new project, and develop a plan to retain and reuse the selected structural elements and components.

[M.6] Design for Disassembly

SCOPE

COMMERCIAL | CORE + SHELL | EDUCATION | HOTELS | LIGHT INDUSTRY | SPORTS

DESCRIPTION

Design building elements and materials for ease of disassembly in order to facilitate future reuse or recycling.

GUIDELINES

Design the building to facilitate future disassembly of components for reuse or recycling to extend the life of materials and conserve natural resources. Building elements and components—including structure, finishing materials, and equipment—should have the ability to be easily separated and disassembled. Measures to facilitate disassembly include the use of modular components, movable partitions, and bolted connections or fastening systems for the structure and/or building envelope. Avoid composite or bonded materials that cannot be separated or reused. Interior finishes and equipment should be designed to be easily removable to allow for refurbishment and remodeling. Select durable and low-maintenance building materials, components, and assemblies, and design for adaptability to ensure longevity.

SPORTS

Sports facilities receive added environmental, cultural, and economic benefit by incorporating flexibility, as well as the ability for disassembly, into the early phase of the design concept by enabling the facility to attract a full range of sports and entertainment events. For larger stadiums, design seating to be flexible with moving tier technology that allows the lower section of the stadium seating to be reconfigured to accommodate events of different scales and spectator capacities. Lightweight, retractable roof systems may be easily reconfigured or removed to provide additional opportunities for use. Component or movable flooring and playing surfaces can provide interchangeable surfaces allowing for multiple sporting events to be conducted in one facility. Partition systems and other temporary structures are beneficial for serving the dissimilar needs of various events. Temporary facilities such as toilets and service equipment can allow the sports facility to serve crowds of varying capacities.

FURTHER RESOURCES

Publications:

1. John, Geraint, et al. *Stadia: A Design and Development Guide*. New York: Elsevier, 2007. Print.

[M.7] Life Cycle Assessment (LCA)

SCOPE

COMMERCIAL | CORE + SHELL | RESIDENTIAL | EDUCATION | MOSQUES | HOTELS | LIGHT INDUSTRY | SPORTS

DESCRIPTION

Encourage the use of materials and products which have the lowest life cycle environmental impact and embodied energy.

GUIDELINES

Embodied energy includes the total amount of energy used to extract, transport, manufacture, distribute, install, maintain and disassemble, deconstruct, or recycle a particular material. Select materials that have a low life cycle environmental impact and embodied energy to minimize overall impact to the environment.

Identify materials that have the lowest Life Cycle Assessment (LCA) values, are available to the region, and are appropriate for the building elements of the proposed project. Evaluate the design options and determine which of these materials can be used for the project. Run budget calculations to set a target for the specific project. These considerations should take place early in the design process to assess which materials will be most appropriate and feasible in terms of the project design and budget.

FURTHER RESOURCES

Publications:

1. Anderson, J., D. Shiers and M. Sinclair. *Green Guide to Specification, X181*. United Kingdom: Construction Industry Research and Industry Association, 2002. Print.

INDOOR ENVIRONMENT [IE]

The majority of a person's day is spent indoors, and the quality of the indoor environment has a direct impact on their health, comfort, well-being, and satisfaction. The building design should promote a comfortable, healthy, and safe environment for building occupants and users. Factors that can improve indoor environmental quality include: monitoring air temperature and quality, as well as adjusting or calibrating as appropriate; maximizing the time period that the building can utilize natural ventilation; designing an adequate mechanical ventilation system; ensuring a sufficient level of illumination, while using an optimal combination of natural and artificial lighting; providing for occupant comfort by minimizing glare; maximizing views to the exterior for all occupants; controlling the amount of noise produced by or transferred from the building interior and exterior; and controlling indoor pollutants and sources of airborne contamination.

Criteria in this category include:

IE.1	Thermal Comfort
IE.2	Natural Ventilation
IE.3	Mechanical Ventilation
IE.4	Illumination Levels
IE.5	Daylight
IE.6	Glare Control
IE.7	Views
IE.8	Acoustic Quality
IE.9	Low-Emitting Materials
IE.10	Indoor Chemical & Pollutant Source Control

[IE.1] Thermal Comfort

SCOPE

COMMERCIAL | CORE + SHELL | EDUCATION | MOSQUES | HOTELS | LIGHT INDUSTRY | SPORTS

DESCRIPTION

Provide a thermally comfortable environment to ensure the comfort and health of building occupants.

GUIDELINES

Provide a thermally comfortable environment with negligible fluctuations in temperature and humidity to ensure the health and comfort of occupants. The thermal comfort of building occupants is dependent upon both environmental conditions and personal factors. Environmental conditions include air temperature, radiant temperature, and relative air velocity and humidity levels, while personal factors include the activity, clothing, and personal preferences of occupants.

Thermal comfort can be attained using one of the following methods: active conditioning (mechanical HVAC systems), passive conditioning (natural ventilation), or a combination of both active and passive conditioning. A hybrid system may be more suitable and effective for projects in a hot, dry climate. The project design should include an efficient and appropriate building envelope to help achieve the desired thermal comfort levels. Use one of the available HVAC load calculation methods to facilitate the selection of appropriate equipment to reach optimal thermal conditions within the building. In the design stage of HVAC system capacity, design the system capacity for a zone satisfying the peak cooling load so that the thermal comfort of occupants in a zone can be guaranteed for the hottest hour.

The following table outlines the measurement type and typical spaces used in the calculation according to the appropriate typology:

TYPOLOGY	MEASUREMENT TYPE	TYPICAL SPACES
COMMERCIAL	PMV	Offices, Reception Areas
CORE + SHELL	PMV	Per Tenant Type
EDUCATION	ADPI	Classrooms, Offices, Special Functional Spaces (e.g. Auditorium)
MOSQUES	ADPI	Prayer Halls
HOTELS	ADPI	Guestrooms
LIGHT INDUSTRY	PMV	Offices, Operational Areas
SPORTS	PMV	Offices

Thermal Comfort Assessment Methods

COMMERCIAL | CORE + SHELL | LIGHT INDUSTRY | SPORTS |

Use the Predicted Mean Vote (PMV) method to assess the thermal comfort of the building's proposed occupied areas. The PMV method evaluates and relates the environmental conditions and personal factors by employing a thermal sensation scale to determine overall thermal comfort. The most important factors for achieving thermal comfort in buildings that are actively conditioned are air temperature and humidity, both of which can be controlled through the design of the air conditioning system. PMV calculations may use the assumed design values for other secondary factors such as air velocity. Consider the mean radiant temperature that can vary due to design decisions such as exterior construction materials and the presence of shading devices. Building simulations may be conducted early in the design process to verify the zoning and control levels necessary to achieve desired thermal comfort levels over the entire year and with variable occupancy/vacancy. Design the HVAC system to be flexible and to respond to part-load demands in order to provide optimal thermal conditions while minimizing energy use.

EDUCATION | MOSQUES | HOTELS

Use the Air Diffusion Performance Index (ADPI) method to assess the thermal comfort of the building's proposed occupied areas. The ADPI method evaluates the spatial conditions of air temperature and speed to convey a single-number means of relating temperatures and velocities in an occupied zone to the occupants' thermal comfort. Higher ADPI values are desirable as they represent a higher comfort level. In order to calculate the ADPI of a space to determine the worst case, measure the temperature and air velocities at points throughout the occupied space. These points are defined by ASHRAE as being the area six feet high from the floor and one foot away from the walls. The temperature and velocity at each measured point are used to calculate an effective draft temperature—a draft temperature of 0 is considered thermally neutral. Negative draft temperatures equate to cooling sensations, while positive values represent a predicted feeling of warmth. For more information, consult the ADPI curves in Chapter 31 of the *ASHRAE Handbook of Fundamentals*.

The most important factors for achieving thermal comfort in buildings that are actively conditioned are air temperature and humidity, both of which can be controlled through the design of the air conditioning system.

EDUCATION

Thermal comfort is essential for a learning environment. Rooms that are too hot or too cold can significantly decrease the attention span and performance of students and teachers. Consider providing each classroom with individual temperature control, allowing teachers to adjust the temperature as necessary. Also, avoid "hot spots" which are extremely bright areas that are created by direct sunlight. Operable windows and user controlled shades and blinds are also recommended.

MOSQUES

A flexible HVAC system is particularly important for prayer areas in mosques, since there is a high fluctuation of occupant density throughout the day.

HOTELS

Guestrooms should have individual thermostats to control the indoor temperature. Smaller hotels most commonly use stand-alone package terminal air conditioners (PTACs) for in-room temperature control. Larger hotels may be better served with a networked system of chillers and boilers that offer localized control. Because studies have shown that many sold guestrooms remain unoccupied for 12 or more hours per day, hotels should link automated systems, such as energy management systems (EMS), reservation systems, and automated check-out systems, so that unoccupied rooms will not use excess energy. Guestroom temperature may be controlled through the EMS, and sold rooms may be adjusted for thermal comfort in advance of guest arrival. In addition, automatic keycard systems can greatly reduce power consumption in unoccupied rooms by shutting off most power-consuming devices like the HVAC unit when a guest leaves and automatically restarting the temperature control and lighting upon return to the room. Providing operable windows in all guestrooms will also enhance the range of thermal comfort options while further reducing energy consumption.

LIGHT INDUSTRY

In light industrial facilities, different consideration should be given to both office and operational areas. Operational areas can require specific heating and cooling requirements such as cold storage areas or painting workshops. When the temperature of certain spaces is above or below normal thresholds, workers should be provided with adequate clothing and protections to ensure their health and comfort. Operational areas also include equipment and processes that have the potential to generate heat which effects building users. Ventilation systems should be designed to handle additional cooling loads in these specific areas.

While office areas associated with light industrial facilities typically have the same requirements for thermal comfort, they are susceptible to heat gains associated with adjacent operational areas. When operational areas generate excessive heat, ventilation systems should be sized to handle the additional cooling loads. Projects should design ventilation systems to handle the cooling and heating loads of both operational and office areas.

SPORTS

For outdoor or semi-outdoor spaces, thermal comfort can be achieved by limiting exposure to the elements. Limiting exposure to extreme solar and heat conditions is an important health and comfort issue, particularly for outdoor spectator seating. Use solar simulations or other analysis to determine the sun angles during the hottest times of the year and layout the spectator seating to avoid those sun angles. The project can also use design features, such as shading devices or a retractable roof, to limit the sun exposure.

FURTHER RESOURCES

Websites:

1. *The Usable Buildings Trust.* Useable Buildings Trust. Web. 30 June 2010. <www.usablebuildings. co.uk/>.

Publications:

1. United States. Environmental Protection Agency. *ENERGY STAR® Building Manual: 12. Facility Type: Hotels and Motels.* Washington: EPA, 2007. Print.

2. *Building Energy and Environmental Modeling, AM11.* United Kingdom: Chartered Institution of Building Services Engineerings, 1998. Print.

3. *Environmental Design, Guide A.* 7th ed. Issue 2. United Kingdom: Chartered Institution of Building Services Engineerings, 2007. Print.

4. *Weather, Solar and Illuminance Data, Guide J.* United Kingdom: Chartered Institution of Building Services Engineerings, 2002.

5. *Trust Heating Control Technology Guide, CTG002.* London: The Carbon Trust, 2006. Print.

6. *Ventilation for Acceptable Indoor Air Quality, ASHRAE Standard 62.1-2004.* Atlanta: American Society of Heating, Refrigeration, and Air Conditioning Engineers, 2004. Print.

7. *Thermal Environmental Conditions for Human Occupancy, ASHRAE Standard 55-2004.* Atlanta: American Society of Heating, Refrigeration, and Air Conditioning Engineers, 2004. Print.

8. Boed, Victor. *Controls and Automation for Facilities Managers: Applications Engineering.* Boca Raton: CRC Press, 1998. Print.

9. Bauman, Fred. *Giving Occupants What They Want: Guidelines for Implementing Personal Environmental Control in Your Building.* Berkeley: Center for the Built Environment, University of California, Berkeley, 1999. Print.

10. Brundett, G.W., L. Harriman, and R. Kittler. *Humidity Control Design Guide for Commercial and Institutional Buildings.* Atlanta: American Society of Heating, Refrigeration, and Air Conditioning Engineers, 2002. Print.

11. Brennan, Terry, James B. Cummings and Joseph W. Lstiburek. "Unplanned Airflows & Moisture Problems." *ASHRAE Journal.* Atlanta: American Society of Heating, Refrigeration, and Air Conditioning Engineers, 2002. Print.

12. Harriman, Lewis and Joseph W. Lstiburek. *The ASHRAE Guide for Buildings in Hot and Humid Climates, Second Edition.* Atlanta: American Society of Heating, Refrigeration, and Air Conditioning Engineers, 2009. Print.

13. *ASHRAE Handbook: Fundamentals.* Washington: American Society of Heating, Refrigerating and Air-Conditioning Engineers, 2009. Print.

[IE.2] Natural Ventilation

SCOPE

COMMERCIAL | CORE + SHELL | RESIDENTIAL | EDUCATION | MOSQUES | HOTELS | LIGHT INDUSTRY | SPORTS

- Rated for SINGLE and GROUP RESIDENTIAL

DESCRIPTION

Encourage effective natural ventilation strategies in conjunction with mechanically ventilated systems.

GUIDELINES

Design the project to effectively utilize natural ventilation strategies within mechanically ventilated buildings to minimize energy use and provide fresh outdoor air to building occupants. Natural ventilation strategies should provide for an adequate flow of air within occupied spaces to ensure appropriate thermal comfort conditions. In buildings employing a naturally ventilated system, it is recommended that the total area of operable windows is at least 5% of the total floor area of the room to provide for effective ventilation. In addition, more than 90% of all occupied spaces should be designed for cross-ventilation.

Design the building for the flexibility to convert between mechanical and natural ventilation strategies as needed. During cooler months, the building may be able to utilize natural ventilation strategies, thereby conserving energy. Additionally, allowing for natural ventilation within a mechanically ventilated building enables fresh air to be introduced into the building, alleviating odors and improving occupant comfort.

Determine the natural ventilation features of the site including the direction of prevailing winds and density of surrounding buildings. Orient the building on the site to take advantage of naturally occurring ventilation patterns throughout the site. Provide operable windows and/or trickle vents on windows to allow for the possibility for natural ventilation in the proposed building. For additional benefits of natural ventilation, use cross-ventilation to promote air flow within the building by placing operable windows on at least two separate walls in one space, preferably on two opposing walls for the most effective air circulation. Ensure that the number, location, operability, and type of windows allow for a sufficient level of fresh air and ventilation. Use calculations to determine effective ventilation within the building, using ventilation design tools recommended by CIBSE AM 10:2005. An analytical model may also be used to predict whether room-by-room airflows will provide adequate ventilation.

The design of the project should allow for users/occupants to open windows at will in order to allow for individual comfort and well-being. Opening mechanisms of all operable windows should be convenient and accessible for ease of use. Operable windows should be located away from sources of pollution to avoid indoor air contamination.

The following table outlines the typical spaces used in the calculation according to the appropriate typology:

TYPOLOGY	TYPICAL SPACES
COMMERCIAL	Office, Reception areas
CORE + SHELL	Per Tenant Type
RESIDENTIAL	Living/family rooms, Bedrooms, Dining rooms
EDUCATION	Classrooms, Libraries
MOSQUES	Prayer Halls
HOTELS	Guestrooms
LIGHT INDUSTRY	Office areas
SPORTS	Indoor competition areas, Fitness suites, Locker facilities, Retail spaces, Offices, Restaurants, Common areas

MOSQUES

Many mosques have courtyards and porticos. These spaces are often shaded and can facilitate natural ventilation of the prayer hall. Allow hot air to escape the prayer hall by placing windows high in the space. Cool air can then enter from the adjacent courtyards or porticos and create effective air circulation.

HOTELS

Hotels with an atrium may also benefit from natural ventilation strategies by taking advantage of the stack effect.

SPORTS

In a sports facility, there is the potential for a number of different functional spaces to be included in the project. Study how these different spaces can use natural ventilation strategies depending on the ventilation requirements of each task to be performed. Sports facilities with retractable roofs frequently do not have the mechanical ventilation systems to handle the ventilation needs when the roof is closed. Projects should take care to implement natural ventilation strategies that can handle the ventilation needs of this type of facility when the roof is closed.

FURTHER RESOURCES

Publications:

1. Abdul Rahman, Abdul Malik and Muna Hanim Abdul Samad. "Natural Ventilation: A Passive Design Strategy in Designing Hotel Lobbies – Cases from Tropical Malaysia." *Archnet-IJAR, International Journal of Architectural Research*. 3.2 (2009). 66-74. Print.

2. *Natural Ventilation in Non Domestic Buildings, AM10*. United Kingdom: Chartered Institution of Building Services Engineers, 2005. Print.

3. *Ventilation and Cooling Option Appraisal - A Client's Guide, GPG 290*. London: The Carbon Trust, 2001. Print.

4. *Natural Ventilation in Non-Domestic Buildings- a Guide for Designers, Developers and Owners, GPG 237*. London: The Carbon Trust, 1998. Print.

5. *Ventilation for Acceptable Indoor Air Quality,* ASHRAE Standard 62.1-2010. Atlanta: American Society of Heating, Refrigerating and Air-Conditioning Engineers, 2010. Print.

[IE.3] Mechanical Ventilation

SCOPE

COMMERCIAL | CORE + SHELL | EDUCATION | MOSQUES | HOTELS | LIGHT INDUSTRY

DESCRIPTION

Provide effective mechanical ventilation to ensure occupant comfort and health.

GUIDELINES

Ensure occupant well-being and comfort by providing effective ventilation for occupants in mechanically ventilated buildings. Ventilation rates and air quality levels should meet the minimum compliance requirements of the following: *ASHRAE Standard 62.1-2004: Ventilation for Acceptable Indoor Air Quality*, the equivalent *CIBSE*, or other accepted standard. An effective ventilation system should promote efficient air exchange and the design should ensure that outdoor air delivered to interior spaces will be successfully delivered to building occupants.

Fresh air intakes should be positioned away from exhaust vents to minimize recirculation. Locate outdoor air intakes away from pollution sources including building exhaust air louvers, exhaust outlets from adjacent buildings, cooling towers, loading docks, air exhaust from waste facilities, parking garages, transportation stops, smoke discharge openings, and dedicated exhausts from toilets and kitchens. Protect outdoor air intake openings from rainwater, animals, and debris with screens and bird guards, and specify ventilation lining that will not release contaminants into the air path. A high-efficiency air filtration system should be utilized to remove particles from outdoor air before being distributed throughout the building. Ensure that the final filter in the system has an expected efficiency of 85-95%.

Minimize risk of contamination by encapsulating or removing exposed insulation inside ducts, air-handling units, and variable-air-volume boxes. Positive building pressurization can be used in hot climates to prevent warm and humid air from seeping into the building.

Provide building occupants/users with the ability to individually control ventilation rates to ensure thermal comfort and well-being. The ventilation air paths of mechanical ventilation systems should be easily accessible for maintenance and servicing. Maintenance plans should provide for regular servicing and verification of the effective operation of all ventilation systems.

When specifying and designing the ventilation systems, consider the balance between fresh air supply and energy efficiency. Carbon dioxide sensors can be used in air-conditioned buildings to ensure appropriate ventilation in response to varying occupancy levels and uses.

MOSQUES

Because occupant density will vary significantly depending on prayer times, HVAC systems should be flexible enough to adapt to various occupancy loads. For example, consider the need to accommodate a large number of people for midday Friday prayers and during the holy month of Ramadan. Other prayer times should have lower occupancy rates and thereby, a lower demand on the ventilation systems.

FURTHER RESOURCES

Publications:

1. *Minimizing Pollution at Air Intakes, CIBSE TM 21.* United Kingdom: Chartered Institution of Building Engineers,1999. Print.

2. United States. Environmental Protection Agency. *Building Air Quality: A Guide for Building Owners and Facility Managers, EPA 402-F-91-102.* Washington: EPA, 1991. Print.

3. Trost, Frederick, and J. Trost. *Efficient Building Design Series, Volume 2: Heating, Ventilating and Air Conditioning.* Upper Saddle River: Prentice Hall, 1998. Print.

4. *Heating, Ventilating, Air Conditioning and Refrigeration, Guide B.* United Kingdom: Chartered Institution of Building Engineers, 2005. Print.

5. *Environmental Design, Guide A.* United Kingdom: Chartered Institution of Building Engineers, 1999. Print.

6. *HVAC Applications Handbook, Chapter 44: Building air intake and exhaust design.* Washington: American Society of Heating, Refrigerating and Air-Conditioning Engineers, 2003. Print.

7. *Best Practice in the Specification of Offices, BCO Guide 2005;* British Council for Offices, 2005. Print.

8. *Ventilation for Acceptable Indoor Air Quality,* ASHRAE Standard 62.1-2004. Washington: American Society of Heating, Refrigerating and Air-Conditioning Engineers, 2004. Print.

9. *Energy Standard for Buildings Except Low-Rise Residential Buildings,* ASHRAE Standard 90.1-2004. Washington: American Society of Heating, Refrigerating and Air-Conditioning Engineers, 2004. Print.

10. *Thermal Environmental Conditions for Human Occupancy,* ASHRAE Standard 55-2004. Washington: American Society of Heating, Refrigerating and Air-Conditioning Engineers, 2004. Print.

11. *Standard Guideline for Using Indoor Carbon Dioxide Concentrations to Valuate Indoor Air Quality and Ventilation,* ASTM D 6245-1998. American Society for Testing and Materials, 1998. Print.

12. Harriman, Lewis G., and Joseph W. Lstiburek. *The ASHRAE Guide for Buildings in Hot and Humid Climates, Second Edition.* Washington: American Society of Heating, Refrigerating and Air-Conditioning Engineers, 2009. Print.

13. United Kingdom: Office of the Deputy Prime Minister. *Approved Document F: Ventilation.* London: United Kingdom Building Regulations, 2006. Web. 06 June 2010. <http://www.gravesham.gov.uk/media/pdf/i/f/ApprovedDocumentF.pdf>.

14. *Energy performance of buildings - Calculation of energy use for space heating and heating,* EN-ISO 13790. Switzerland: International Organization for Standardization, 2008. Print.

[IE.4] Illumination Levels

SCOPE

COMMERCIAL | CORE + SHELL | EDUCATION | MOSQUES | HOTELS | LIGHT INDUSTRY | SPORTS

DESCRIPTION

Ensure light levels that have been designed in line with best practices for visual performance and comfort.

GUIDELINES

Design the project's lighting systems to ensure adequate illumination levels and light quality for the visual comfort and well-being of building occupants. Reduce the energy needed for electrical lighting by providing daylighting, being mindful that introducing daylight into building interiors may also increase solar heat gain and cooling loads. Ensure through calculations and simulations that benefits obtained from daylighting will not be compromised by a significant increase in energy use. The quality and quantity of daylighting within the building is affected by window placement and sizes, glazing transmittance, room geometry, interior surface finishes, and shadows cast from nearby buildings. Daylight can be filtered into interior spaces through the use of plants, blinds, shades, light-scattering glazing types, or architectural devices such as louvers and baffles. Consider the use of skylights, roof monitors, and clerestories to allow light into interior spaces, especially at the core of the building.

Provide light levels no less than those recommended in the *IESNA Lighting Handbook, 9th Edition*. Determine the appropriate light levels for each of the different task-related spaces in the proposed building and design the lighting system to meet these requirements. For example, tasks that take place in a conference room are considered 'visual tasks of high contrast and large size,' and would require a horizontal illuminance value of 30 foot-candles according to the IESNA standard. Avoid over-illumination of entire rooms or spaces by providing individual task or accent lighting where higher illumination levels are needed.

Lighting systems should allow for adjustments by the occupants as necessary to promote comfort and maximize visual performance. Lighting controllability is the amount of control an individual has to turn lights off and on, adjust brightness, and change the positioning of fixtures. The extent and type of lighting controls should relate to the function of each space, the amount of occupants, the frequency of use, and the level of daylighting within each space. Coordinate light fixture layouts in conjunction with furniture layouts to maximize lighting efficiency.

The following important factors should be considered when specifying fixtures to improve light quality and visual comfort within the building: luminance ratio limits, veiling reflections, reflected glare, shadows, color, and intensity. Minimize illumination intensity and lighting power through the specification and selection of energy-efficient fixtures that use efficacious sources. Use high frequency fittings to minimize discomfort due to the flicker caused by luminaires that have a low frequency, such as conventional fluorescent luminaries. Consider lighting design strategies including light shelves and indirect lighting systems that reflect light off the ceiling, providing uniform and ambient lighting conditions. The ceiling shape may be designed to more efficiently distribute light from windows or skylights, such as through the use of a sloped or curved ceiling. The color, texture, and reflectance of surface materials in a room can also help improve lighting conditions and minimize lighting needs.

The following table outlines the typical spaces used in the calculation according to the appropriate typology:

TYPOLOGY	TYPICAL SPACES
COMMERCIAL	Offices, Reception Areas
CORE + SHELL	Per Tenant Type
EDUCATION	Classrooms, Office
MOSQUES	Prayer Halls
HOTELS	Guestrooms, Offices
LIGHT INDUSTRY	Offices, Operational Areas
SPORTS	Indoor Competition Areas, Fitness Suites, Locker Facilities, Retail Spaces, Offices, Restaurants, Common Areas

EDUCATION

The wide range of visually intensive activities that take place within an educational facility require a range of lighting techniques. Some of these techniques are outlined below.

Classrooms
The variety of activities that take place in a classroom require a highly flexible lighting system that is user controlled. A recommended method for electric lighting is the use of suspended linear lights above a flat, white ceiling. This system works best in spaces that have a ceiling height of at least 2.75 meters. The electric lights should be integrated with daylighting strategies. One method for doing this is with daylight sensors, allowing electric lights to be automatically dimmed or switched off when there is sufficient daylight. Some lighting fixtures are multi-scene, which allows them to switch between direct and indirect lighting (downward or reflected off the ceiling). This is particularly useful in classrooms and spaces that use projectors and require a darkened room. For classrooms where dry-erase boards are used often, consider providing lighting specifically for the board. Dry-erase boards have more contrast than chalkboards. Therefore, they require less lighting; however, they are highly reflective. Place electric lighting carefully to avoid glare. Avoid placing dry-erase boards on a wall that is opposite a daylighting source.

Corridor Lighting
Recessed fluorescent luminaires or surface mounted corridor luminaires are recommended. These lights should have daylight sensors if they will be integrated with a daylighting strategy. Ensure that the lighting

has an emergency back-up system. Choose fixtures that are durable since hallways are more prone to vandalism.

Gym Lighting

Industrial high bay fluorescent luminaires with T-5HO or T-8 lamps are recommended. Another option is to use multiple compact fluorescent luminaires in a single housing. Be sure to protect the lamps from flying balls.

Library Lighting

It is recommended that spaces used for reading have 20 - 50 footcandles. As for classrooms, suspended fluorescent luminaires are a good choice for spaces that have a ceiling height of at least 2.75 m. Task lighting should be used as well. Table lamps with compact fluorescent bulbs are an energy efficient option. Under-shelf task lights using T-8 or modern T-5 lamps are another option.

MOSQUES

Light levels in prayer halls should be comfortable for reading purposes. Avoid over-illumination of entire rooms or spaces by providing individual task or accent lighting, where higher illumination levels are needed.

HOTELS

Many zones in a hotel, such as public areas, back-of-house service areas, and office areas, may contain light sources that operate continuously over a 24-hour period. As a result, lighting plays a significant role in energy consumption in a hotel. Maximizing available daylight can greatly reduce the lighting needs in common spaces and guestrooms. For deep-plan hotels, consider designing atria with wells featuring proper geometric angles and reflective surfaces to bring daylight into the core of the building.

Hotel typologies may vary greatly and include a variety of differently programmed spaces including guestrooms, dining areas, recreational facilities, etc. Therefore, all spaces should provide light levels no less than those recommended in the *IESNA Lighting Handbook, 9th Edition*. Furthermore, ASHRAE standards mandate that all hotel and motel guestrooms and suites must have a master control device at the main room entry that controls all permanently installed luminaires and switched receptacles. Master controls for lighting may also be integrated into the hotel's electronic key card system to guarantee automatic shut-off when guestrooms are unoccupied. Also consider the use of automatic lighting control systems for spaces where both daylight and artificial lighting are used to minimize energy use (e.g. atria, indoor pools, retail areas, etc.). Automatic lighting control systems include the use of timers, daylight controls, and occupancy/vacancy controls. Sensors can be used within the lighting system to adjust illumination levels in response to variable daylighting conditions and user occupancy/vacancy. Provide dimmable and stepped daylighting controls for greater control over the building's electric lighting system.

SPORTS

For sports facilities, there are often requirements for illumination levels related to the individual sport or competition. These requirements can dictate the lighting levels, glare, intensity, or color of the light, in addition to other specific requirements. Refer to the individual technical documents and guidelines for each sport and organization to determine the appropriate lighting layouts. Consider the many different users of a sports facility, including competitors, spectators, and support staff, as well as the different lighting needs for each type of functional space. Perform calculations or simulations to ensure that each user and functional component meets the appropriate lighting levels.

FURTHER RESOURCES

Websites:

1. *Radiance WWW Server.* Lawrence Berkeley National Laboratory. Web. 28 June 2010. <http://radsite.lbl.gov/radiance/>.

2. *Whole Building Design Guide.* National Institute of Building Sciences. Web. 28 June 2010. <http://www.wbdg.org/design/index.php>.

Publications:

1. *Recommended Practice of Daylighting, RP-5-99.* New York: Illuminating Engineering Society of North America, 1999. Print.

2. Rea, Mark S., ed. *The IESNA Lighting Handbook.* 9th ed. New York: Illuminating Engineering Society of North America, 2000. Print.

3. *American National Standard Practice for Office Lighting, RP-1-04.* New York: Illuminating Engineering Society, 2004. Print.

4. *Code for Lighting: Part 2.* United Kingdom: Chartered Institution of Building Services Engineers, 2004. Print.

5. *Office Lighting, Lighting Guide 7.* United Kingdom: Chartered Institution of Building Services Engineers, 2005. Print.

6. Boed, Viktor. *Controls and Automation for Facilities Managers: Applications Engineering.* Boca Raton: CRC Press, 1998. Print.

7. Steffy, Gary. *Architectural Lighting, Second Edition.* New York: Jon Wiley and Sons, Inc., 2002. Print.

8. Guzowski, Mary. *Daylighting for Sustainable Design.* New York: McGraw-Hill, 1999. Print.

9. Ander, Gregg D. *Daylighting Performance and Design, Second Edition.* John Wiley & Sons, 2003. Print.

10. *Daylighting and Window Design, Lighting Guide 10.* United Kingdom: Chartered Institution of Building Services Engineers, 1999. Print.

11. United Kingdom. *Lighting for Buildings: Code of Practice for Daylighting, BS 8206-2.* London: British Standards Institution, 2008. Print.

12. Rea, Mark S., ed. *The IESNA Lighting Handbook.* 9th ed. New York: Illuminating Engineering Society of North America, 2000. Print.

13. Switzerland. *Football Stadiums: Technical Recommendations and Requirements.* Switzerland: Federation Internationale de Football Association, 2004. Web. 10 October 2010. <http://www.fifa.com/mm/document/tournament/competition/football_stadiums_technical_recommendations_and_requirements_en_8211.pdf>.

[IE.5] Daylight

SCOPE

COMMERCIAL | CORE + SHELL | RESIDENTIAL | EDUCATION | MOSQUES | HOTELS | LIGHT INDUSTRY | SPORTS

• Rated for SINGLE and GROUP RESIDENTIAL

DESCRIPTION

Optimize the exposure of daylight for interior spaces in order to improve light quality for building occupants and reduce the need for artificial lighting.

GUIDELINES

Incorporate strategies for daylighting to increase comfort and well-being of the users and minimize energy consumption from artificial lighting. Integrate daylighting into the overall lighting approach of the building to provide a balance between natural and artificial lighting. Determine the lighting needs in the spaces throughout the development and take measures to maximize the daylighting potential of the building.

Incorporate daylighting strategies through building orientation and the design of exterior and interior spaces. Orient the project away from obstructing objects and other buildings to capture the maximum amount of light. Consider design elements, such as atria, courtyards, skylights, and shading devices, to harvest and control natural light. Deep-plan hotels can bring daylight in through atria and use angled wells and reflective surfaces to bring daylight into the core of the building to illuminate lobbies and interior circulation spaces. Design indoor recreational spaces, such as pools or fitness rooms, with large expanses of glazing and skylights to maximize daylight infiltration.

Incorporate large window openings in areas of maximum daylight exposure. Minimize depth of rooms and building floor plates to increase the amount of natural light entering the space. Increase the quantity of natural light by promoting design elements such as light shelves, light ducts, and other apparatus to capture light.

Design the project to balance and control factors such as heat gain and loss, glare, visual quality, and variations in daylight availability. Specify low reflective interior color schemes and materials to balance visual quality and quantity. Consider the use of sun shades, louvers, operable blinds and draperies, and exterior light shelves to control and reduce glare. Design frit patterns for glazing surfaces and specify glass which has the ability to reduce solar heat gain while allowing natural light into the space.

Use technology to design and implement integrated daylighting strategies throughout the project. Consider computer modeling software to simulate daylighting conditions and develop an effective strategy for both natural and artificial light throughout the project. Other technologies such as photo responsive controls can be used to maintain consistent light levels and minimize the change in the quality of light from natural to artificial.

Daylight can be modulated in a variety of ways, making it an appropriate solution for the different spaces within a building. The following outlines some of these daylighting techniques:

View Window: A view window is vertical glazing at eye level. While this can be a good source of daylight, there is the potential for glare and for "hot spots" created by direct sunlight. Louvers, blinds, and shades can help prevent these negative effects.

High Side-lighting/Clerestory with Light Shelves: As with view windows, this technique brings in daylight from the side. However, placing the window high on the wall helps to evenly distribute the light through the room. The use of a light shelf will also help with light distribution and can help minimize glare. High side-lighting works best in spaces with high ceilings.

Wall Wash/Top-lighting: This technique is achieved by placing a skylight next to a wall. The daylight gets distributed over the surface of the wall, providing indirect light.

Central Top-lighting: Louvers or baffles should be used to avoid direct sunlight and to evenly distribute the light.

Patterned Top-lighting: This technique consists of multiple skylights spread out over a large area in a grid. It produces even, low glare illumination which is good for large areas like gymnasiums, libraries, and cafeterias. Good spacing for the skylight grid is approximately 1 to 1.5 times the floor to ceiling height.

Linear Top-lighting: This is a good technique for hallways. It produces a strong linear light that can aid in orientation.

Tubular Skylights: These skylights reflect daylight down through tubes. If they are used in a grid of proper dimensions, they can provide an even distribution of light. They are good for retrofits and useful in areas with deep roof cavities, since they can be designed to fit between the framing of a roof.

The following table outlines the typical spaces used in the calculation according to the appropriate typology:

TYPOLOGY	TYPICAL SPACES
COMMERCIAL	Offices, Reception Areas
CORE + SHELL	Per Tenant Type
RESIDENTIAL	Living/Family Rooms, Bedrooms, Dining Rooms
EDUCATION	Classrooms, Libraries, Offices
MOSQUES	Prayer Halls
HOTELS	Guestrooms, Offices
LIGHT INDUSTRY	Offices, Operational Areas
SPORTS	Indoor Competition Areas, Fitness Suites, Locker Facilities, Retail Spaces, Offices, Restaurants, Common Areas

EDUCATION

The use of controlled daylight in an educational facility can improve comfort and performance. Daylight can be modulated in a variety of ways, making it an appropriate solution for many different kinds of spaces within an educational facility. Employ proper daylighting techniques to provide adequate illumination for classroom tasks without causing glare or distraction.

MOSQUES

Prayer halls should have enough illumination to allow for occupants to read comfortably. Because prayer halls have high ceilings, high side-lighting works best for overall illumination. When using view windows, special consideration should be taken for prayer halls where direct views to the outside may be deemed a distraction to worshippers.

FURTHER RESOURCES

Publications:

1. Baker, Nick and Koen Steemers. *Daylight Design of Buildings.* London: James & James, 2002. Print.

2. Phillips, Derek. *Daylighting: Natural Light in Architecture.* Oxford: Elsevier, 2004. Print.

3. Rea, Mark S., ed. *The IESNA Lighting Handbook.* 9th ed. New York: Illuminating Engineering Society of North America, 2000. Print.

4. Rutes, Walter A. *Hotel Design: Planning and Development.* New York: W.W. Norton & Company, 2001. Print.

5. *High Performance Schools Best Practices Manual, Volume 2, Design.* San Francisco: The Collaborative for High Performance Schools, 2006. Web. 06 July 2010. <http://www.chps.net/dev/Drupal/node/288>.

[IE.6] Glare Control

SCOPE

COMMERCIAL | CORE + SHELL | EDUCATION | LIGHT INDUSTRY | SPORTS

DESCRIPTION

Minimize direct or reflected glare within occupied spaces to improve visual comfort for occupants.

GUIDELINES

An excessive amount of luminance within the visual field produces glare, which can cause discomfort to building occupants. The issue of glare is exacerbated when there are high-contrast situations, which typically occur when the illuminance from a window opening is significantly higher than the luminance of adjacent surfaces that are dark and low-reflectance. Design for low luminance ratios and appropriate lighting of interior surfaces to maximize occupant comfort. Provide adequate controls to prevent glare within the building, especially in regularly occupied spaces. All projects will avoid excessive daylight glare, as per CIBSE LG7:2001 & BSEN12464-1:2002 Standards. Particular care should be taken to prevent glare in spaces where occupants use visual display terminals (VDT).

Measures to block direct sunlight and reflections from bright external surfaces through windows, glazed doors, and skylights include light shelves, blinds, louvers, fins, shades, tinted glazing, and light-scattering glazing. All controls should be convenient and easily accessible for building occupants to maximize comfort. The design of the project may also provide mechanically operated shading devices with manual override capabilities. Consider the spatial arrangement of the building and interior spaces to minimize discomfort from excessive glare and contrast.

Windows and other openings that bring in daylight can produce more severe glare than artificial lighting because of the intensity of sunlight. However, appropriate glare control for artificial lighting is also important for occupant comfort and the project should specify luminaries that meet the glare and illuminance requirements set forth in *IESNA Lighting Handbook, 9th Edition*.

Refer to IESNA RP-5-99 for further guidelines to reduce glare.

The following table outlines the typical spaces used in the calculation according to the appropriate typology:

TYPOLOGY	TYPICAL SPACES
COMMERCIAL	Office, Reception areas
CORE + SHELL	Per Tenant Type
EDUCATION	Classrooms, Libraries, Offices
LIGHT INDUSTRY	Office, Operational areas
SPORTS	Indoor competition areas, Offices

SPORTS

Adhere to the applicable design and lighting guidelines for each sport to meet the appropriate glare standards. Glare is most likely to occur when a person's view angle is in the spill light of a lighting fixture and not in the direct light. For a large space, such as an enclosed competition area, arrange the light fixtures to account for the view angles between the players, spectators, and the lighting fixtures. Ensure even illumination throughout the space as well as even displacement of the lighting fixtures. Reduce the potential for glare by laying out an even array of lighting fixtures. Also, limit the tilt angle of each lighting fixture to reduce the amount of glare produced.

FURTHER RESOURCES

Publications:

1. *Recommended Practice of Daylighting, RP-5-99.* New York: Illuminating Engineering Society of North America, 1999. Print.

2. Rea, Mark S., ed. *The IESNA Lighting Handbook.* 9th ed. New York: Illuminating Engineering Society of North America, 2000. Print.

3. *Daylighting and Window Design, Lighting Guide 10.* United Kingdom: Chartered Institution of Building Services Engineers, 1999. Print.

4. *The Visual Environment for Display Screen Use, Lighting Guide 3.* United Kingdom: Chartered Institution of Building Services Engineers, 1996. Print.

5. *Light and lighting. Lighting of work places. Indoor work places*, BSEN12464-1: 2002. London: British Standards Institution, 2003.

6. *Lighting Guide 7: Office Lighting*, CIBSE LG7: 2005. United Kingdom: Chartered Institution of Building Services Engineers, 2005. Print.

[IE.7] Views

SCOPE

COMMERCIAL | CORE + SHELL | EDUCATION | SPORTS

DESCRIPTION

Provide occupants with access to external views.

GUIDELINES

Configure the building form, interior spaces, and partitions in such a way to maximize views to the surrounding environment. A view of the outdoors is important for the comfort and well-being of occupants as it allows the eyes to refocus, helps to relieve eye strain, and reduces visual fatigue. Design to provide views from spaces that are regularly occupied; it is not necessary to consider those spaces that are intermittently used throughout the day such as storage rooms, copy rooms, circulation spaces, restrooms, and mechanical and custodial spaces. Use interior glazing, an open plan configuration, and lower partition heights to maximize views to the exterior. Consider arranging the plan so that spaces in frequent use are along the exterior of the building, while infrequently used spaces, including restrooms, elevator cores, and storage rooms, are contained within the core of the building. The size of the windows, type of glazing, and placement of shading devices can impact the ability for occupants to have views out to the exterior. Ensure that views out are not impeded by the design of the façade and glazing types.

Views to the surrounding environment should provide a view of a landscape or distant buildings if possible, to allow occupants to focus on objects rather than solely a view to the sky. The site plan should be analyzed to determine whether the proximity of buildings on adjacent sites will impede views to the exterior. Designing for views into an internal courtyard may also be considered as a view to the exterior, provided that the courtyard is of an appropriate depth.

Providing views to the exterior directly correlates with the amount of glazing or openings in the building, which may increase solar heat gain and increase cooling loads. Ensure through calculations and simulations that benefits obtained from maximizing views and daylighting will not be compromised by a significant increase in energy use.

FURTHER RESOURCES

Publications:

1. Bickford, E. Lawrence, O.D. "Computers and Eyestrain." *The EyeCare Reports.* Santa Barbara: Larry Bickford,1996. Web. 04 August 2010. <http://www.eyecarecontacts.com/abstracts_and_reports_home.html>.

2. *Daylighting and Window Design, Lighting Guide 10;* CIBSE: Chartered Institution of Building Services Engineers. United Kingdom: Chartered Institution of Building Services Engineers, 1999. Print.

[IE.8] Acoustic Quality

SCOPE

COMMERCIAL | CORE + SHELL | RESIDENTIAL | EDUCATION | MOSQUES | HOTELS | LIGHT INDUSTRY | SPORTS

- Rated for SINGLE and GROUP RESIDENTIAL

DESCRIPTION

Meet minimum requirements for acoustic quality within the building.

GUIDELINES

Design the project to meet minimum requirements for acoustic quality within the building to ensure a satisfactory level of acoustic performance. Interior noise levels should be maintained so as not to interfere with the regular tasks of occupants and users. Consider noise generated both outside the building, on the project's site or adjacent properties, as well as noise generated from services, equipment, and activities within the proposed building. Determine the specific requirements for each space, including privacy levels, sound isolation needs, and acceptable background noise levels. Design for the appropriate acoustic performance levels within each area of the building.

The design of the project should mitigate the effects of external noise sources through the layout of the building, its placement on the site, and the location of noise-sensitive spaces within the building plan. Use vegetation, earth berms, or other noise barriers on the site as a means of muffling off-site noise before it reaches the building. Specify building components with appropriate sound transmission class (STC) rating, such as exterior walls, windows, and doors, to protect interior spaces from harmful noise sources. STC is a number rating system used to compare the sound insulation properties of building elements including walls, floors, ceilings, windows, and doors. STC ratings are impacted by the height of partition walls; in cases where sound isolation and/or privacy levels are important, specify full-height partition constructions. Consider using acoustical ceiling tiles and wall panels or spray-on acoustical treatments in spaces where additional sound absorption is necessary.

Provide sufficient noise insulation to mitigate impacts from interior noise sources such as those generated by plumbing systems, mechanical ventilation systems, and air conditioning equipment. Minimize excessive vibration from services and equipment as per the *ISO 2631-2:1989* standard in order to mitigate acoustic problems in the building interior. Separate noise-generating areas from noise-sensitive spaces within the building. Consider the use of soft, sound absorbent materials for interior finishes including walls, floors, and ceilings in order to reduce noise levels. A higher sound absorption rate will attenuate noise transferred from the exterior or generated within the building, and will increase the acoustic performance within the building. Acoustical information can be obtained from manufacturers to help select the most appropriate and effective materials and components to meet the project's acoustical requirements.

Consider the floor impact sound level and the performance of sound insulation as related to impact noises both heavy and light. An example of light floor impact noise is a chair being dragged on a concrete floor, whereas a heavy floor impact noise might be the sound of children jumping. Sound insulation performance varies with the type of structural system, floor, and ceiling construction methods, slab types and thicknesses, flooring materials, and the size of interior spaces. Impact insulation class (IIC) measures the impact of sound insulation of a floor/ceiling construction. Specify the appropriate IIC in the design of the building construction to provide the proper acoustic performance levels for interior spaces.

Background Noise: The indoor noise level must meet the recommended noise level for speech intelligibility as described in *BS 8233:1999* for the typical occupied spaces listed in the following table:

Typical Occupied Space	Design Range $L_{Aeq,T}$ dB	
	Good	Reasonable
Private Office	40	50
Open-plan Office	45	50
Living Rooms	≤ 40	
Bedrooms	≤ 30	
Classroom	30	40
Prayer Halls	30	35
Guestrooms	30	35

The following table outlines the typical spaces used in the calculation according to the appropriate typology:

TYPOLOGY	TYPICAL SPACES
COMMERCIAL	Offices, Reception Areas
CORE + SHELL	Per Tenant Type
RESIDENTIAL	Living/Family Rooms, Bedrooms, Dining Rooms
EDUCATION	Classrooms, Offices, Special Functional Spaces (e.g. Auditoriums, Assembly Halls)
MOSQUES	Prayer Halls
HOTELS	Guestrooms
LIGHT INDUSTRY	Office, Operational Areas
SPORTS	Offices

EDUCATION

Appropriate acoustic conditions in education buildings is essential to learning. Students are still developing language ability through their teens, and noise affects this development. A study has shown that children are much more affected by background noise than adults since their ability to suppress echo-like sounds is still developing (Litovsky, 1997). In addition, noise causes teachers to speak more loudly and strain their voice.

Some of the common sources of noise in classrooms are traffic, HVAC systems, reflective surfaces, and hallways. The design of large rooms which are divided into smaller classrooms with partitions should be avoided, as this usually produces spaces with poor acoustics.

Program placement can have a large effect on acoustic quality. Locate noisy areas like gymnasiums and cafeterias away from more noise-sensitive spaces such as classrooms and libraries. Partition walls should extend to the structural deck, rather than ending at the suspended ceiling. Windows are often the weakest noise barriers in a building. Double glazed windows are much better at insulating a space from noise. Consider laminated glass for the outer panes of double glazed windows. Many acoustic panels and tiles have reduced acoustical properties if they are painted.

The standard reverberation time (RT) for classrooms is between 0.4 and 0.6 seconds. An RT of greater than 0.6 seconds causes echo and makes speech less intelligible. However, if too much of the sound is absorbed (generally an RT of less than 0.4 seconds), then a speaker's voice may not carry far enough. This is particularly a problem in larger classrooms and lecture halls. Reverberation times should be considered for all spaces. The recommended maximum reverberation time for auditoriums is 0.8 seconds. The recommended maximum for gymnasiums is 1.2 seconds.

MOSQUES

Special consideration should be given to the design of prayer halls to enhance acoustic quality, particularly during prayer services. Worshippers in all prayer halls, including the main prayer hall and women's prayer area, should be able to clearly hear the Imam during congregational prayers.

HOTELS

In each hotel suite or room, install decibel level regulators on all appliances in addition to mechanical equipment such as HVAC. Consider installing sound buffeted windows to reduce traffic noise. Apart from rooms and suites, hotel facilities should remain below an overall decibel level in order to prevent distraction to hotel occupants. Kitchen noise, large laundry machines, and gym equipment provide the largest amount of in-hotel noise pollution on average; restrict each to night time operation when possible to avoid occupant disturbance. Special consideration should be given to soundproofing guestrooms that are situated below or above banquet rooms, restaurant/lounge spaces, sports facilities, and other areas with high noise levels. In addition, hotels near airports or busy highways will need to implement additional measures to regulate external noise transmission.

SPORTS

Provide sufficient noise insulation to mitigate impacts from sporting events. For sports facilities, there are often requirements for acoustic conditions related to the individual sport or competition. These requirements can dictate the acoustic conditions, reverberation time, or other specific needs for that space. Refer to the individual technical documents and guidelines for each sport and organization to determine the appropriate acoustic conditions that must be met. Consider the many different users of a sports facility, including competitors, spectators, and support staff, as well as the different acoustic requirements for each type of functional space. Perform calculations or simulations to ensure that each user and functional component meets the appropriate acoustic conditions.

FURTHER RESOURCES

Publications:

1. *Sound Insulation and Noise Reduction for Buildings- Code of Practice, BS8233.* British Standards Institution, 1999. Print.

2. United Kingdom. *Acoustics- Measurement of Sound Insulation in Buildings and of Building Elements, BS EN ISO 140-4.* British Standards Institution, 1998. Print.

3. United Kingdom. *Acoustics- Rating of Insulation in Buildings and of Building Elements, Part 1 and 2, BS EN ISO 717-1, 717-2.* British Standards Institution, 1997. Print.

4. United Kingdom. Office of the Deputy Prime Minister. *Resistance to the Passage of Sound, Building Regulations Approved Document E.* London: United Kingdom Building Regulations, 2003. Print.

5. *Mechanical vibration and shock -- Evaluation of human exposure to whole-body vibration - Part 2: Vibration in buildings (1 Hz to 80 Hz),* ISO 2631-2-1989. Switzerland: International Organization for Standardization, 1989. Print.

6. *Procedure for Estimating Occupied Space Sound Levels in the Application of Air Terminals and Air Outlets,* ARI Standard 885-1998. USA: Air-Conditioning and Refrigeration Institute, 1998.

7. Switzerland. *Football Stadiums: Technical Recommendations and Requirements.* Switzerland: Federation Internationale de Football Association, 2004. Web. 10 October 2010. <http://www.fifa.com/mm/document/tournament/competition/football_stadiums_technical_recommendations_and_requirements_en_8211.pdf>.

[IE.9] Low-Emitting Materials

SCOPE

COMMERCIAL | CORE + SHELL | RESIDENTIAL | EDUCATION | MOSQUES | HOTELS | LIGHT INDUSTRY | SPORTS

- Rated for SINGLE and GROUP RESIDENTIAL

DESCRIPTION

Meet minimum emissions targets for indoor materials and finishes to ensure the comfort and health of occupants.

GUIDELINES

Indoor air contaminants are harmful to the environment and to the health and comfort of building occupants. Minimize the health risks associated with indoor air contaminants by selecting indoor materials that have zero or minimal rates of volatile organic compound (VOC) emissions. Indoor materials to be considered are the following: paints, coatings, primers, finishes, stains, sealants, caulking, adhesives, carpets, and composite wood products. Avoid composite wood and agrifiber products that contain urea-formaldehyde resins. Any material that can affect the indoor air quality through its emissions should be selected carefully taking into account its potential for harmful emissions. Materials used for any indoor surfaces, such as flooring, walls, and ceilings and those used within wall cavities, above suspended ceilings, and below finished floors, should have low VOC emissions. Materials used for indoor furnishings, mechanical system components, and any other systems or components within the building that may emit harmful contaminants into the indoor environment should be avoided.

Water based paint usually has a lower rate of VOCs than solvent based paint, since most of the VOC emissions are from evaporating solvents. Low VOC paint is generally considered to be paint with a VOC content of less than 100 mg/L. Lighter color paints tend to have a lower VOC content. Make sure that paint is applied and dries before soft materials (carpet, etc.) that easily absorb VOC content are present.

Wood composite materials (particle board, medium density fiberboard, etc.) often use high VOC adhesives. Many alternatives are available including recycled plastic, salvaged wood, oriented strand board (OSB), and certified wood. Biocomposites are also available, however they should not be used in moisture prone areas. Low density fiberboard, a good option for tack boards, is made from 100% recycled paper. Specify the smallest amount of adhesive possible. Also, when possible, have boards cut off-site in a space with adequate ventilation.

Product manufacturers can provide Material Safety Data Sheets (MSDS) to help assess the potential health risks associated with certain materials. MSDSs are not completely comprehensive in terms of impacts on

indoor air quality, but include information on relevant factors such as: hazardous components, chemical identification, chemical characteristics, reactivity data, health hazard data, fire/explosion hazard data, etc. Additionally, acquire emissions test data from manufacturers for each indoor material and finish. Reports should include information on material origins, test methods and results, and VOC emission rates.

FURTHER RESOURCES

Websites:

1. *Commercial Adhesives (GS-36)*. Washington: Green Seal Standards and Certification, 2000. Web. 3 June 2010. <http://www.greenseal.org/certification/standards/commercial_adhesives_GS_36.cfm>.

2. *Carpet and Rug Institute*. Carpet and Rug Institute. Web. 3 June 2010. <www.carpet-rug.org>.

Publications:

1. *An Update on Formaldehyde*. Consumer Product Safety Commission. Web. 3 June 2010. <http://www.cpsc.gov/CPSCpub/pubs/725.html>.

2. *Green Seal Environmental Standard for Paints and Coatings (GS-11)*. 2nd ed. Washington: Green Seal, Inc., 2008. Web. 3 June 2010. <http://www.greenseal.org/certification/standards/commercial_adhesives_GS_36.cfm>.

3. *High Performance Schools Best Practices Manual, Volume 2, Design*, CHPS [The Collaborative for High Performance Schools], 2006

4. *Rule #1168: Adhesive and Sealant Application*. Diamond Bar, CA: South Coast Air Quality Management District, 2005. Web. 3 June 2010. <http://www.aqmd.gov/rules/reg/reg11/r1168.pdf>.

[IE.10] Indoor Chemical & Pollutant Source Control

SCOPE

COMMERCIAL | CORE + SHELL | EDUCATION | HOTELS | LIGHT INDUSTRY | SPORTS

DESCRIPTION

Minimize potentially hazardous airborne contaminants affecting building occupants.

GUIDELINES

Minimize the risk to human health and comfort by reducing hazardous particulates and chemical/biological contaminants in the indoor air. Physically isolate areas that may generate harmful contaminants from the mixing and storage of chemicals, such as maintenance and custodial spaces. High-volume copier, fax, and printer operations should also be isolated from other occupied spaces as these activities can generate harmful contaminants. Physically separate adjacent spaces from areas with potential contaminants through the use of deck-to-deck partitions or sealed gypsum board enclosures. Use dedicated exhaust systems in these areas and utilize negative pressurization to prevent contaminants from entering adjacent spaces and the building's main ventilation systems.

Pedestrian traffic into the building can introduce contaminants in the form of debris, dirt, and dust. The project design should incorporate an entryway system with grilles, grates, or other effective systems to capture potentially harmful particles as occupants enter the building. Select and design entryway systems with a recessed floor area for collecting debris and particles as users enter the building. Systems with a recessed floor area allow for easier maintenance and cleaning and are more effective than other systems such as carpeted entryways.

Air handling units that process both return air and outside supply air should utilize high-level filtration systems. Select air handling units in part for their capacity to accommodate required filter sizes and pressure drops. Ensure easy access to air handling units for regular servicing and maintenance.

Prevent the growth of fungus, mold, and bacteria on building surfaces through the project design and selection of appropriate indoor materials. Indoor materials should resist microbial growth; consider specifying hard surfaces, such as tile and wood flooring rather than carpets, as they are easier to maintain and clean. Where carpets are to be used, plan for regular high performance cleaning and replacement as necessary. Design the building envelope with effective moisture barriers to prevent water from seeping into the building and to protect interior materials and finishes. Minimize the risk of Legionella through the proper design and location of wet cooling towers and building water treatment systems. Legionella bacteria can move into the building through outside air intakes and pose a serious health risk to building occupants.

Maintain water quality in water cooling towers using means such as chemical injection and automatic blowing, and ensure that high temperatures in water heaters are maintained to discourage bacterial reproduction.

HOTELS

Hotels bring together people from international locations and a variety of workplaces, and as a result, visitors can unwittingly bring harmful substances with them to the hotel. It is important to minimize the risk to human health and comfort by reducing hazardous particulates and chemical/biological contaminants in the indoor air. Because hotels are responsible for extensive maintenance and continuous cleaning procedures, extra precaution should be taken to ensure proper storage of cleaners and other chemicals. Harmful substances should be securely stored in back-of-house areas away from guests and separated according to usage type. In order to prevent accidental reactions, avoid storing pool chemicals with other volatile substances. Enforce housekeeping procedures that minimize guest exposure to cleaning agents and other chemicals.

LIGHT INDUSTRY

In light industrial facilities, any areas where contaminants and sources of pollution are handled must be isolated and properly ventilated. Provide ventilation systems to ensure that the air quality levels in operational areas meet or exceed acceptable levels for worker health and safety. Use exhaust equipment and containment measures in close proximity to equipment and processes to reduce the amount of contamination that enters the spaces. Ensure that ductwork and other equipment handling potentially contaminated air is properly sealed or isolated. Additionally, provide workers with appropriate safety equipment including masks or respirators.

Operational areas in light industrial facilities must be isolated from offices and other support areas because those spaces are not usually designed to handle indoor contaminants. Walls and windows should be sealed properly to avoid air leakage between the two spaces. Keep office areas at a positive pressure to further reduce infiltration. Mechanical systems that support the two areas should also be separated to avoid any contaminated air exhaust mixing with the fresh air intakes.

SPORTS

Sports facilities may include spaces with potential contaminants such as swimming pools. While these facilities are a necessary part of the facility, additional measures should be provided to ventilate and control those contaminants to ensure the health and safety of building users. When possible, specify chemicals and other maintenance materials with low VOC contents.

FURTHER RESOURCES

Websites:

1. *Green Seal; The Mark of Environmental Responsibility.* Green Seal, n.d. Web. 04 August 2010. <http://www.greenseal.org/index.cfm>.

2. *Environmentally Preferable Purchasing.* EPA, 12 May 2010. Web. 04 August 2010. <http://www.epa.gov/opptintr/epp/>.

Publications:

1. *Method of Testing General Ventilation Air-Cleaning Devices for Removal Efficiency by Particle Size,* ANSI/ASHRAE Standard 52.2-1999. Atlanta: American Society of Heating, Refrigerating and Air-Conditioning Engineers, 1999. Print.

2. Health and Safety Executive. *Legionnaires' Disease; The Control of Legionella Bacteria in Water Systems. The Approved Code of Practice and Guidance.* 3rd ed. Surrey, UK: HSE Books, 2000. Print.

3. *Minimising the Risk of Legionnaires Disease, CIBSE TM 13.* United Kingdom: Chartered Institution of Building Services Engineers, 2002. Print.

4. *Minimizing the Risk of Legionellosis Associated with Building Water Systems,* ASHRAE Standard 12-2000. Atlanta: American Society of Heating, Refrigerating and Air-Conditioning Engineers, 2000. Print.

5. Bennett, K.M. *Efficient Humidification in Buildings, Application Guide 10/94.1.* United Kingdom: Building Services Research and Information Association, 1995. Print.

CULTURAL & ECONOMIC VALUE [CE]

Maintain the region's cultural identity and contribute to economic growth through the design and planning of the new building. The building design should align with cultural identity and traditions, integrating the building into the existing cultural fabric. In addition, specify local materials and a local workforce to encourage the support of the national economy.

Criteria in this category include:

CE.1	Heritage & Cultural Identity
CE.2	Support of National Economy

[CE.1] Heritage & Cultural Identity

SCOPE

COMMERCIAL | CORE + SHELL | RESIDENTIAL | EDUCATION | MOSQUES | HOTELS | SPORTS

• Rated for SINGLE and GROUP RESIDENTIAL

DESCRIPTION

Encourage design expression that will align with and strengthen cultural identity and traditions.

GUIDELINES

Qatar has experienced tremendous development in the past several decades. With the rapid growth and change of Qatar's cities and landscapes, the country must address how to preserve its cultural resources and define the role of these resources in the formation of a cultural and national identity. The conservation of heritage in the region is not a new pursuit. In the spring of 1985, the Arab Urban Development Institute (AUDI) worked with the Arab Towns Organization (ATO) and the Union of Municipalities of Marmara Region to discuss many of the pressing questions facing Qatar today. During "The Conference on the Preservation of Architectural Heritage of Islamic Cities," participants raised questions about city planning, historic preservation, and new architecture and development. The conference addressed methods of conservation and also focused on the question of cultural and national identity.

While past studies have laid much of the conservation framework for addressing architectural and urban scale, it is important to learn from limitations of prescriptive preservation systems. Preservation is the process of maintaining living contact with the past through the identification, transmission and protection of that which is considered culturally valuable and therefore worthy of retaining. Often, the emphasis is on formal concerns, such as architectural style, and the result can be superficial. Central to any system of cultural preservation should be an understanding of the plurality of the culture within the region and the identification of cultural values. Not only will cities have culturally distinct regions and neighborhoods, but there will be architectural diversity within individual districts. Complexity and diversity within a culture underlines the need for the careful review on a case by case basis, rather than a restrictive overarching approach. According to the Aga Khan, "The loss of [this] inheritance of cultural pluralism—the identity it conveys to members of diverse societies, and the originality it represents and stimulates in all of them—will impoverish our societies now and into the future."

A common method of preserving cultural identity in some cities is through the creation of "historic districts." At best, these protected areas are highly successful livable neighborhoods; in the worst cases, they can resemble "reconstructed villages"—protected enclaves that draw much tourism but ultimately are little more

than theme parks. Urban designers now agree that cities, like living organisms, grow incrementally and are most successful when their growth is diachronic, including both old and new. It is equally important for any urban plan to be flexible and adaptable to accommodate change. It is the fine balance of preserving the old and encouraging the new, as well as the traditional and the innovative, that makes urban life socially, economically and aesthetically vital while maintaining cultural continuity and prosperity.

URBAN LEVEL

Urban Form and Structure

This includes the street pattern and types of blocks. The scale, alignment, and hierarchy of buildings should be considered—for example, the use of compact, urban form. It is also important to study areas of enclosure and convergence. Both form and structure may be affected by the area's natural topography.

Uses and Patterns of Activity

This is the range and nature of activity. The commercial and residential tradition and the use of the ground floors is very important in adding a cyclical dimension and variety to the neighborhood.

Neighborhood Dimension

Consider pedestrian and vehicular traffic and the scale of experience. Factors to consider include the width of streets and sidewalks, street furniture, landscaping, and whether vehicular traffic should be allowed.

Open Space

What is the role of open/public space? Where does it exist in the city/development? Where do people gather and interact?

Archaeological Sensitivity

Are there any landmark buildings on the site which are historically protected or need historical protection? Is the site archaeologically important? If so, determine the allowable extent of new construction.

Urban Response to Natural Environment

How is wind and sunlight regulated at an urban scale? Consider the width and orientation of streets in relation to the direction of prevailing winds and solar position. Shade can be provided through the use of narrow streets and strategically placed buildings and vegetation.

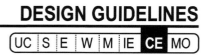
BUILDING LEVEL

Facades

Consider the rhythm of buildings, windows, and openings. This should influence the design and composition of the façade and other elements along the street.

Height

How does the building relate to the height of other buildings?

Scale

How does the building relate to the scale of other buildings? This is related to but distinct from height as it is relative to the size of the human body and experience of the user. It can be affected by proportion, massing, composition, and the setback from the street. Also, consider the scale and proportion of building elements (arches, columns, etc.) and how they relate to each other as well as the composition and space of the building as a whole.

Lighting

How will the building/development appear at night? Consider the intensity, placement, and color of light. Also consider what aspects of the architecture are to be highlighted.

Style

How does the building relate to the architectural style of the region and surrounding buildings? It does not need to replicate the style; however, there should be a sensitive response to the style of the existing architectural context.

Material/Color

Consider the material of buildings and surfaces. How will the material age, and how will this affect its relationship to the site? How does the building's color relate to its context? Using local materials and trades is preferable to maintain both cultural and economic value.

Space

Consider using open space such as courts and *liwans* (galleries). How do these spaces relate to more private spaces and to the entry sequence?

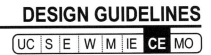
Building Response to Natural Environment

How is wind and sunlight regulated at the building scale? The orientation and placement of buildings in relation to the direction of prevailing winds and solar position greatly influences how they will react to the natural environment. Ways to regulate the environment include natural ventilation, the use of courtyards, and strategic shading of the building.

INNOVATION

Develop contemporary design using the basis of the Qatari vernacular architecture.

APPENDIX

Architectural Context

The style of the architectural heritage in Qatar is similar to what is seen in other Gulf areas. The Qatari, or Gulf Style, is a hallmark of the identity of the place known as the Gulf, which specifically includes the countries that borders the Arabian Gulf to the east or west, Iran (or what is known as Fariss), and countries west of the Arabian Peninsula's shores bordering the Gulf.

The Gulf Style is distinct and has characteristics and an identity that cannot be mixed with the styles that surround it. For example, venturing deep into the Arabian Peninsula (the area of Najd specifically), one starts to observe differences in style, construction materials, climate, and culture. This change also occurs in Iran, where a different culture and style emerges as one travels inland from the Persian shore.

The architectural style in Qatar and around the Gulf basin is the result of a group of influences, factors, and forces that shaped the Gulf Style over many years. These agents can be divided into two realms: social, cultural, religious, or spiritual influences and climatic, geological nature, or materialistic components.

After centuries of development, the Gulf Style attained a distinct identity during the middle of the twentieth century. In the mid-twentieth century, new influences emerged due to the economic boom from oil production in the area, as well as the arrival of architects and contractors from different cultures that introduced new construction technologies and materials such as cement.

The character and the cultural, social, and religious values of the Arabic-Muslim community in the Gulf have contributed to the space planning and layout of the typical house. For example, *Al-Majlis*, where the owner of the house would meet his guests, is an important component that provides the guests with hospitality and lodging in this space without compromising the privacy of his family. *Al-Majlis* is located near the entry and separated from the house by a zigzagging corridor (named *Dehleez*) that serves to block the guests from the yard of the house while still allowing the guests to freely enter and exit the space, as needed.

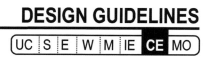

Climate in the Gulf Area

At the Tropic of Cancer, the climate is dry and humid depending on the time of year, and the wind direction most of the year is North-Western, delivering heat and humidity during the summer. The summer breeze (*Al-Barih*) blows during May and June, hot and dry with dust. The Autumn season (known as *Al-Sfari* season) starts in October, bringing rain (*Al-Wasmi*). Winter follows with a sharp drop in temperature and cooler northerly winds (*Shamal*).

The seasonal climate has clearly influenced the planning of the typical house. The orientation of the house and the openings are designed to attract the desired breeze, while protecting the house from the hot breeze (*Al-Barih*) and the dust in the summer, or the cold northerly winds in the winter. Climatic awareness is equally shared among the Gulf residents and users of the buildings, as well as the architects and constructors who produce them. The exploitation of the positive climatic conditions, while protecting areas from the negative climatic effects, resulted in highly efficient space planning, and the lack of electricity was not a hindrance in making the region habitable.

Geological Nature and Building Materials

The geological nature and properties of the building materials in this part of the world greatly affected the formal language of the buildings. For centuries, architects and masons honored the practice of building with local stone and mud mortar, and they transferred the technological expertise to future generations of builders. Due to the properties of stone and mortar, builders could not build walls more than one meter high in a single day. Instead, they had to wait for the mortar to dry and then proceed higher in daily increments.

Later, column and beam technology emerged, saving both time and construction materials. This method clearly expressed the structural nature of the columns and beams, allowing them to remain visible on the interior and exterior of the building. The column (50 cm x 50 cm) was the optimum size for stone and mud infill, but the use of horizontal beams (*Jussur*) was essential to stabilize the columns and prevent the walls from collapsing. After reaching a height of two meters, the mason would place wood beams, that at the time were imported from east Africa, to horizontally brace the wall. This process was repeated up the wall and a top cross-beam helped stabilize the roof. The spaces between the columns were usually between 90 cm -100 cm and had different uses depending upon the homeowner's budget.

The structural form and the nature of the construction materials lends a unique aesthetic to the architecture of Qatar. Therefore, the Qatari Gulf Architectural Style can be called "structural architecture" because the columns and beams are clearly visible and define the structural plan. The interpretation of this style in many new works has resulted in unintended misrepresentation and distortion, due to a lack of accurate scientific references for architects and designers.

Gypsum used in plaster is expensive to mine and import, therefore it was used according to the owner's budget. Sometimes it was used in important spaces, such as *Al-Majlis* (the male guests' reception area) and the entryway of the house, or it was used for the decoration of columns and parapets. The elements of decoration and ornamentation in Qatari Gulf architecture are distinct and have features that cannot be confused with the arts of the surrounding areas. Nevertheless, the scarcity of scientific examples needed to create an elaborate and objective reference of ornamental motifs in the Gulf Style often results in confusion and cultural misappropriation.

Qatari Gulf architecture is absent of color except for the color resulting from the gypsum plaster on the walls of the house. The coloration that is seen is the result of the interaction of the gypsum with the natural elements; its color turns ivory after one year and will darken more over time from the dust.

Despite the lack of variations in color, the façades of the buildings are enriched with alternating recesses and protrusions. The shaded verandah, or *Al-Laywan,* is recessed deep (about three meters) from the exterior sunlight and creates a very dark shadow. The walls built between the columns are recessed about ten centimeters from the face of the columns, casting shadows that create elegant lines over the building façades.

MOSQUES

Mosques in Qatar have historically had simple designs. Influenced largely by climate, the mosques are characterized by thick walls that minimize heat gain from the sun. Courtyards provide a comfortable space outdoors and many prayer halls have entryways shaded with a portico. A common feature of traditional Qatari mosques is the open iwan, or external prayer hall, usually located between an inner iwan and a courtyard. The entrance to the mosque is typically through the courtyard, with the minaret at one corner of the site plan.

Consider community involvement in the design of the mosque. A mosque expresses a collective identity that is tied to both the local community as well as the global community of Islam. Input from the congregation can help in the development of a mosque design that appropriately expresses the community and region.

FURTHER RESOURCES

Publications:

1. Jaidah, Ibrahim and Malika Bourennane. *The History of Qatari Architecture 1800-1950.* Milan: Skira, 2010. Print.

2. Kahera, Akel, Latif Abdulmalik, and Craig Anz. *Design Criteria for Mosques and Islamic Centers: Art, Architecture, and Worship.* Oxford: Elsevier, 2009. Print.

3. Holod, Renata and Hasan-Uddin Khan. *The Contemporary Mosque: Architects, Clients and Designs since the 1950s.* New York: Rizzoli, 1997. Print.

4. Al-Wakil, Abdul-Wahed, Adel. "The Principles of Traditional Design of Mosques." *Proceedings of the Symposium on Mosque Architecture. Volume 10: Invited Papers.* Eds. Dr. Mohammed Eben Abdullah Eben Saleh and Dr. Abdelhafeez Feda Alkokani. Riyadh, Saudi Arabia: College of Architecture and Planning, King Saud University, 1999. 1 - 5. Print.

[CE.2] Support of National Economy

SCOPE

COMMERCIAL | CORE + SHELL | RESIDENTIAL | EDUCATION | MOSQUES | HOTELS | LIGHT INDUSTRY | SPORTS

- Rated for SINGLE and GROUP RESIDENTIAL

DESCRIPTION

Maximize the percentage of construction expenditures for goods and services originating from the national economy.

GUIDELINES

Procure construction-related products and services from within the region in order to support the local economy. Investigate the regional availability of products and services, and develop a plan to utilize and employ local companies and firms where possible. The proposed project should consider the enduring viability of existing businesses in the region, provide business spaces within the building where needed, and conduct studies to identify opportunities for investment in the new building or development.

Projects should plan to obtain commitment from contractors to use local labor during the construction phase. The use of local labor minimizes the need for temporary foreign laborer housing, which can reduce the overall cost of construction. Also, the use of local labor maximizes the national employment rate. Additionally, projects should consider the possibility of generating jobs for the local region from the new building or development to promote and support the regional economy.

SPORTS

Development of new sports facilities creates long-term construction related to additional infrastructure and multiple facilities to support the spectators, athletes, and competition venues. Sports facilities utilize construction products, such as steel and concrete, at such a large scale that the effects on the local, national, and global economy should be considered.

Beyond the construction phase, sports facilities create local jobs during the operations phase for peak and legacy events. Additionally, sports facilities create an influx of jobs from the parking attendants, janitorial staff, restaurant and retail employees, to office managers, and media personnel. Sports facilities can provide additional opportunities for investment and development by providing a venue for alternate events, such as concerts, and conducting studies to identify previously successful investments that thrive in proximity to development of sports facilities, such as sporting goods stores and restaurants.

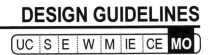
MANAGEMENT & OPERATIONS [MO]

The project should plan for and implement sustainable and effective building management and operations practices. Effective building management and operations can mitigate environmental impacts such as water depletion, materials depletion, and human comfort and health. Factors to consider include the following: creation of a commissioning plan; managing organic waste and recycling practices; detection of major leaks; monitoring and managing energy and water use; managing the major building systems; planning for hospitality services; educating building users on the project's sustainable features; and designing for the building's legacy. Planning for the efficient management and operations of the proposed building will reduce costs associated with energy use and maintenance and ensure occupant comfort, well-being, and satisfaction.

Criteria in this category include:

MO.1	Commissioning Plan
MO.2	Organic Waste Management
MO.3	Recycling Management
MO.4	Leak Detection
MO.5	Energy & Water Use Sub-metering
MO.6	Automated Control Systems
MO.7	Hospitality Management Plan
MO.8	Sustainability Education and Awareness Plan
MO.9	Building Legacy

[MO.1] Commissioning Plan

SCOPE

COMMERCIAL | CORE + SHELL | RESIDENTIAL | EDUCATION | MOSQUES | HOTELS | LIGHT INDUSTRY | SPORTS

- Rated for GROUP RESIDENTIAL

DESCRIPTION

Encourage commissioning planning within the design process.

GUIDELINES

Appoint an independent Commissioning Authority to develop a Commissioning Plan for the project to ensure efficient design, construction, calibration, and performance of all building systems as required by code and local building regulations. A successful commissioning process is dependent upon many factors including proper planning, thorough documentation, effective coordination and communication, and strategic implementation. The Commissioning Authority will be responsible for leading the commissioning process, coordinating with the project team and at times reviewing documentation to ensure proper implementation.

The Commissioning Authority should consult with the project owner to identify the systems to be covered through commissioning and to establish project goals for each of the building systems. Requirements and goals should be quantifiable and measurable in order to easily verify whether the project objectives have been achieved.

The Commissioning Plan typically covers all the critical building systems such as the following: life safety systems, HVAC systems, lighting systems and controls, electrical systems, building envelope, water-use systems, and renewable energy systems. The Commissioning Plan will ensure optimum performance of all relevant building systems under projected occupancy loads and conditions. Furthermore, the Commissioning Authority should develop a plan to address the coordination between team members of all phases, including design, installation, and operation. Coordination between phases is necessary to maintain the performance of the building's systems at the maximum efficiency throughout the life of the building.

The commissioning process should address the performance criteria for each system. The Commissioning Authority will review the necessary documents, such as design documents, submittals, and field testing reports, to verify that the building systems are properly designed and installed to perform efficiently. Additionally, the Commissioning Authority will verify that the commissioning requirements have been included within the construction documents and project specifications. Finally, the Commissioning Authority

will review, verify, and compile the performance results of the building systems and complete a summary commissioning report to document the results of the commissioning process.

HOTELS

Hotel specific mechanical systems, such as in-room HVAC and card reading door locks, should be considered by the Commissioning Authority to ensure proper installation and care of these systems.

FURTHER RESOURCES

Websites:

1. *Cx Assistant Commissioning Tool.* Energy Design Resources, 2005. Web. 30 June 2010. <http://www.ctg-net.com/edr2002/cx/>.

Publications:

1. Armstrong, J., and G. T. Machin. *Boilers: CIBSE Commissioning Code B.* London: Chartered Institution of Building Services Engineers, 2002. Print.

2. *Guideline 0-2005 - The Commissioning Process.* Atlanta: American Society of Heating, Refrigerating and Air-Conditioning Engineers, 2007. Print.

3. *Guideline 1-1996 -The HVAC Commissioning Process.* Atlanta: American Society of Heating, Refrigerating and Air-Conditioning Engineers, 1996. Print.

4. *HVAC & R Technical Requirements for the Commissioning Process.* Atlanta: American Society of Heating, Refrigerating and Air-Conditioning Engineers, 1996. Print.

5. *Measurement, Testing, Adjusting, and Balancing of Building HVAC Systems.* Atlanta: American Society of Heating, Refrigerating and Air-Conditioning Engineers, 2008. Print.

6. Burkhead, Carl E. *Guidance for the Preparation of Operations Plans.* Topeka: Department, 1993. Print.

7. Butcher, Ken. *Commissioning Code M: Commissioning Management.* London: Chartered Institution of Building Services Engineers, 2003. Print.

8. *Commissioning Code L: Lighting.* London: Chartered Institution of Building Services Engineers, 2003. Print.

9. *Commissioning Code W: Water Distribution Systems.* London: Chartered Institution of Building Services Engineers, 2003. Print.

10. Marcus, Dicks. *Commissioning Management: How to Achieve a Totally Functioning Building.* Berkshire, UK: Building Services Research and Information Association, 2002. Print.

11. Heinz, John A., and Richard B. Casault. *The Building Commissioning Handbook.* 2nd ed. Alexandria: Association of Higher Education Facilities Officers, 2004. Print.

12. *Installation and Commissioning of Refrigeration Systems, GPG 347.* London: The Carbon Trust, 2003. Print.

13. Parsloe, C., and A. W. Spencer. *Commissioning of Pipework Systems: Design Considerations.* Berkshire, UK: Building Services Research and Information Association, 1996. Print.

14. Parsloe, C. *Commissioning of HVAC Systems, Technical Memoranda 1/88.1.* Berkshire, UK: Building Services Research and Information Association, 2002. Print.

15. -----. *Commissioning of VAV Systems in Buildings, AG 1/91.* Berkshire, UK: Building Services Research and Information Association, 1991. Print.

16. -----. *Commissioning of Water Systems in Buildings, AG 2/89.2.* Berkshire, UK: Building Services Research and Information Association, 2002. Print.

17. -----. *Commissioning Water Systems Application Principles, AG 2 89.3.* Berkshire, UK: Building Services Research and Information Association, 2002. Print.

18. -----. *Pre-Commission Cleaning of Pipework Systems, AG 1/2001.1.* Berkshire, UK: Building Services Research and Information Association, 2001. Print.

19. -----. *Pre-commission Cleaning of Pipework Systems (2nd Edition): Including Advice on Fit-out Works.* Bracknell, Berkshire: BSRIA, 2004. Print.

20. -----. *Commissioning Air Systems. Application Procedures for Buildings (AG 3/89.3 (2001)).* Bracknell, Berkshire: BSRIA, 2001. Print.

21. Pennycook, Kevin. *Automatic Controls: CIBSE Commissioning Code C.* London: Chartered Institution of Building Services Engineers, 2001. Print.

22. Teekaram, Arnold, and Anu Palmer. *Variable Flow Water Systems: Design, Installation and Commissioning Guidance, AG 16/2002.* Berkshire, UK: Building Services Research and Information Association, 2002. Print.

23. Welch, Terry, and Ken Butcher. *Refrigerating Systems: CIBSE Commissioning Code R.* London: Chartered Institution of Building Services Engineers, 2002. Print.

24. Wilson, J. *Commissioning Code A: Air Distribution Systems.* London: Chartered Institution of Building Services Engineers, 1996. Print.

[MO.2] Organic Waste Management

SCOPE

COMMERCIAL | CORE + SHELL | RESIDENTIAL | EDUCATION | HOTELS | LIGHT INDUSTRY | SPORTS

- Rated for GROUP RESIDENTIAL

DESCRIPTION

Encourage planning for the collection and composting of organic waste in order to minimize waste taken to landfills or incineration facilities.

GUIDELINES

Develop and implement an Organic Waste Management Plan to collect, store, compost, and/or recycle organic waste generated during building operations. Organic waste is material that comes from plant or animal sources that contain carbon compounds. Organic waste is more biodegradable than inorganic materials, and the by-products produced after organic waste has been broken down can be used for composting and enriching the soil. If feasible, calculate the anticipated amounts of organic waste to be generated by the building, in order to provide sufficient space for sorting and storage.

Provide sufficient collection points for organic waste throughout the building, especially near concessions and other food service locations where the majority of organic waste is produced. Collection bins should be provided along with other recyclable or trash containers to ensure building users can easily separate organic waste. Consider specifying self-closing airtight systems in areas containing organic waste to prevent risk to human health. These systems could be designed to be operated either automatically or manually, depending on user preference and their intended use.

In addition to collection points located throughout the facility, provide a central sorting and storage area for organic materials. The sorting and storage areas should be properly contained and ventilated to avoid the dispersion of noxious fumes and odors into occupied spaces of the building, which could present a possible health risk or discomfort to building occupants and users. Locate the organic waste storage spaces in close proximity to vehicular access to facilitate collection and removal.

Organic waste can be composted for reuse as fertilizer for use on- or off-site. Determine the amount of organic waste the project will generate, as well as the amount of compost that can be reused on-site for landscaping or agriculture. Determine the types of organic waste that will be produced, including food waste, yard trimmings, or wood waste, and how these types of organic materials can be used. Identify off-site facilities where excess organic waste can be transported. These off-site locations can be municipal facilities that handle and distribute large quantities of organic material, as well as other smaller facilities that can reuse the material themselves. Ensure that all of the organic waste material that is generated and collected can be used either on- or off-site. The project may also consider using the biomass of generated organic waste as energy. Organic waste generates heat as it is broken down and this energy can be harnessed to provide heat and power for the building.

In addition to organic waste reuse, consider recycling kitchen generated cooking grease. This grease is not easily disposed of in sewers or landfills and can instead be reused for various processes. For example, vegetable-based kitchen grease can be used in biodiesel-run machines, or it can be used as raw materials for the rendering industry.

HOTELS

Because hotels generate a significant amount of organic waste, include reuse options in the Organic Waste Management Plan in order to further diminish the overall organic waste load. Use nutrient rich food debris from hotel food service operations to create compost that improves on-site soil conditions.

FURTHER RESOURCES

Websites:

1. United States. Environmental Protection Agency. "Reducing & Recycling." *Wastes*. US, 24 March 2010. Web. 05 August 2010. <http://www.epa.gov/osw/conserve/materials/organics/reduce.htm>.

2. California. City of Roseville. "How to Recycle Kitchen Grease." *City of Roseville California*. n.d. Web. 05 August 2010. <http://www.roseville.ca.us/civica/filebank/blobdloadasp?BlobID=12619#page=>.

3. New York. New York State Department of Environmental Conservation. "Composting of Organic Waste." 2011. Web. 12 April 2011. < http://www.dec.ny.gov/chemical/8798.html>.

4. United States. Environmental Protection Agency. "Common Wastes and Materials: Organic Materials." 16 December 2010. Web. 12 April 2011. <http://www.epa.gov/osw/conserve/materials/organics/>.

Publications:

1. Diaz, Luis F., Clarence G. Golueke, George M. Savaage, and Linda L. Eggeth. *Composting and Recycling Municipal Solid Waste*. Boca Raton: Lewish Publishers,1993. Print.

2. Zaman, Atiq Uz. "Life Cycle Environmental Assessment of Municipal Solid Waste to Energy Technologies." *Global Journal of Environmental Research* 3.3 (2009): 155-163. Print.

3. Connecticut. State of CT. Dept of Environmental Protection. Bureau of Waste Management. Division of Planning and Standards. *Best Management Practices for Grass Clipping Management*. Hartford: Bureau of Waste Management, 1999. Print.

4. New York. The City of New York Department of Energy Conservation. *City of New York Comprehensive Solid Waste Management Plan*. New York, NY: DSNY, 2006. Print.

[MO.3] Recycling Management

SCOPE

COMMERCIAL | CORE + SHELL | RESIDENTIAL | EDUCATION | MOSQUES | HOTELS | LIGHT INDUSTRY | SPORTS

- Rated for GROUP RESIDENTIAL

DESCRIPTION

Encourage space planning to designate containment facilities for the building's recyclable waste streams in order to minimize waste taken to landfills or incineration facilities.

GUIDELINES

Develop a Recycling Management Plan for the collection, storage, and removal of recycling generated during the building's operations in order to reduce the amount of waste taken to landfills or for incineration. Provide sufficient collection points for recyclable materials throughout the facility, especially near concessions and other food service locations where the majority of waste is produced. Collection bins should be provided, along with other organic or trash containers, to ensure building users can easily separate recyclables. Depending on the recycling capabilities of the region, provide infrastructure to recycle materials such as glass, plastics, paper, cardboard, and metals. Clearly label all collection facilities and provide easy access to facilitate recycling. Signage within recycling facilities should separate the bins for various materials to avoid contamination and improper sorting.

In addition to collection points located throughout the facility, provide a central sorting and storage area for recyclable materials. The collection and storage space should be clearly labeled for recycling, easily accessible to building occupants and facility operators and situated in close proximity to vehicular access to facilitate collection and transport. Consider the impact on the building's indoor environmental quality in terms of unwanted odors and disruptive noises associated with the sorting and storage of recyclable materials. Storage facilities should be properly contained and ventilated to minimize negative impacts to the surrounding building spaces. Recycling activities, such as sorting and hauling, should take place before or after normal business hours to minimize disruption to building occupants and users. Evaluate possible security measures in cases where recyclable materials may be of a high value.

Ensure that storage capacity will be sufficient for the anticipated amount of recyclable materials generated during normal building operations. Consider the size of equipment and facilities to be used for recycling management, such as compactors and wheeled bins, when allocating and designing the collection and storage spaces. Consider various recycling management equipment and strategies, including recycling chutes, compactors, balers, and individual collection bins, located throughout the building to promote and encourage recycling activities. Instruct building occupants and operators on appropriate recycling procedures to maximize recycling rates. Consider activities to reduce materials needed to be disposed of or recycled such as through the use of reusable bags and bottles.

Recycling infrastructure may not be readily available in the region, but the project may still provide recycling facilities to encourage the future consideration and development of such infrastructure in the local community and beyond. Together with the building owner, designers should determine the specific requirements and goals of the project and design with future provisions in mind.

MOSQUES

Consider recycling equipment placement away from 'clean' areas such as post-purification worship spaces and ablution rooms. Recycling collection should take place at times that do not disrupt worship services.

FURTHER RESOURCES

Publications:

1. Adler, David. *Metric Handbook: Planning and Design Data, 2nd ed.* Oxford: Architectural Press, 1999. Print.

2. Diaz, Luis F., Clarence G. Golveke, George Savage, Linda Eggerth. *Composting and Recycling Municipal Solid Waste.* Boca Raton: CRC Press, 1993. Print.

3. *Guide G: Public Health Engineering.* London: Chartered Institution of Building Services Engineers. 2004. Print.

4. Lund, Herbert F. *The McGraw-Hill Recycling Handbook.* New York: McGraw Hill, 2000. Print.

5. Winter, John P., Ann Marie. Alonso, and Gina Goldstein. *Waste at Work: Prevention Strategies for the Bottom Line.* New York: INFORM, 1999. Print.

[MO.4] Leak Detection

SCOPE

COMMERCIAL | CORE + SHELL | EDUCATION | MOSQUES | HOTELS | LIGHT INDUSTRY | SPORTS

DESCRIPTION

Minimize the impact of major water leakages by installing leak detection systems.

GUIDELINES

Develop a Leak Detection Plan to install and maintain an effective leak detection system for all water supply and wet areas in the project, including the building and within the site boundary. Wet areas are defined as any area within a building connected to a water supply system including bathrooms, showers, laundry facilities, and sanitary compartments. The leak detection system must be capable of detecting major leaks, thereby reducing the impact on water consumption and depletion.

The system should be activated when higher than normal flow rates are detected at water meters, for longer than a pre-set time period. The leak detection system should be clearly audible when activated to ensure that building facility operators are alerted of major leaks. Leak detection systems should allow for a high degree of sensitivity to identify various levels of leakage rates. The system's settings should be programmable and adjustable to allow for the appropriate and required values according to the building's water consumption targets. Select a leak detection system that minimizes the likelihood of false alarms that may occur in systems, such as chillers, that consume large amounts of water during standard operations.

FURTHER RESOURCES

Publications:

1. Australia. Department of Water. *Guidelines for Water Meter Installation.* Government of Western Australia, 2007. Print.

2. Lahlou, Zacharia. "Leak Detection and Water Loss Control." *Tech Brief: A National Drinking Water Clearinghouse Fact Sheet.* New York: National Drinking Water Clearinghouse, 2001. Web. 05 July 10. <http://www.nesc.wvu.edu/ndwc/pdf/OT/TB/TB_LeakDetection.pdf>.

3. Mays, Larry W. *Water Distribution Systems Handbook.* New York: McGraw-Hill, 2000. Print.

4. Satterfield, Zane, PE, and Vipin Bhardwaj. "*Water Meters. National Environmental Services Center."* New York: National Drinking Water Clearinghouse, 2004. Web. 05 July 10. <http://www. nesc.wvu.edu/ndwc/pdf/OT/TB/TB_LeakDetection.pdf>.

[MO.5] Energy & Water Use Sub-metering

SCOPE

COMMERCIAL | CORE + SHELL | EDUCATION | HOTELS | LIGHT INDUSTRY | SPORTS

DESCRIPTION

Encourage the installation of energy and water sub-meters that separately monitor systems for in-use energy and water consumption.

GUIDELINES

Develop an Energy and Water Use Sub-metering Plan to install sub-metering devices to monitor and evaluate energy and water performance and consumption during building operations. Major energy and water systems should be metered and monitored in conjunction with data logging to provide for continued accountability of energy and water consumption over the building's lifespan. Energy and water sub-metering will also facilitate the development of strategies to help improve performance, thereby ensuring the overall efficiency of building operations. Target performance values should be set for building energy and water consumption, and facility operators should develop and implement plans to reach and maintain those values. Monitoring devices should display and record the energy and water consumption data of major systems in the building. Energy sub-meters must be provided for all major energy-consuming systems, such as lighting, hot water heaters, boilers, fans, cooling, humidification, space heating, competition-related equipment, large-scale broadcast and media systems, equipment associated with industrial processes, and large-scale food service equipment, to be used in the building. Water sub-meters must be provided for all major water-consuming systems, such as bathroom fixtures, hot water heaters, boilers, chilled water systems, competition-related equipment, and large-scale food service equipment.

Energy and water sub-metering should be properly and clearly labeled, easily accessible, and convenient for regular access by the building facility operators. Specify the appropriate location of energy and water sub-meters such as in the plant room, distribution room, or control room. Determine the optimal quantity and specific locations of energy and water meters according to the types and complexity of systems to be monitored. Consider utilizing energy simulations or engineering analysis to predict overall energy consumption and evaluate major energy system performance. Similarly, predict the overall water consumption to evaluate the efficiency of water-consuming systems. Determine measures and strategies for continued improvement of energy and water efficient building operations, throughout the design of the project and during building operations.

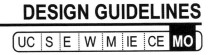

FURTHER RESOURCES

Websites:

1. *Efficiency Valuation Organization.* Efficiency Valuation Organization, 2010. Web 30 June 2010. <http://www.evo-world.org/>.

Publications:

1. *Building Energy Metering – A Guide to Energy Sub-metering in Non-domestic Buildings.* London: Chartered Institution of Building Services Engineers, 2006. Print.

[MO.6] Automated Control System

SCOPE

COMMERCIAL | CORE + SHELL | EDUCATION | MOSQUES | HOTELS | LIGHT INDUSTRY | SPORTS

DESCRIPTION

Encourage the installation of an Automated Control System to optimize system performance.

GUIDELINES

Install an Automated Control System (ACS) that controls and monitors major building systems including cooling, lighting, ventilation, and irrigation. The implementation of an ACS Plan should be customized to work with the specific installed systems in a building to create a central interface of data collection and systems control. An ACS consists of a variety of components and can significantly reduce energy consumption and increase occupant comfort by integrating multiple systems and optimizing performance. Another benefit of an ACS is the recorded data available to building managers. This data is essential for calibration and optimization and can alert the building manager to critical maintenance issues. The location of the ACS control panel should be secured and easily accessible to the building manager. It is important that there is sufficient space near the ACS for necessary wiring and cables. Consider providing space for new systems to be added as technology advances.

Daylight control may be achieved by installing a Digital Addressable Lighting Interface (DALI). This system gives each light unit its own individual digital address, which allows for software-based, network control of luminaries. When coupled with structured wiring that combines power and data cabling in a connectable modular system, a DALI system makes rapid physical reconfiguration of lighting systems possible.

Since lighting systems consume a significant amount of energy, the ACS may also employ daylight harvesting technologies to manage energy usage and occupant comfort levels. With a careful balance of natural light and artificial fluorescent luminaries, a building can benefit from considerable energy savings as well as an improvement of spatial quality. Install sensors to monitor and automatically adjust lighting levels to favor the usage of natural light, and make sure artificial lights have dimmers to respond to these adjustments. To optimize lighting, it is necessary that daylight sensors cover an appropriate area. If a daylight sensor is covering an area that is too large, it will not provide enough detailed information to ensure proper lighting levels. Since the perimeter of a building will often have more varied conditions than internal areas, it is important to carefully consider and target zones.

Fade time for lighting sensors is critical to maintain occupant comfort, with very slow fade times being the least disruptive. Select appropriate shading systems, such as blinds and louvers, to work with the automated control and place sensors in appropriate locations relevant to these systems. It is important that lights and blinds have user overrides and that building occupants are educated about the ACS.

Multifunction sensors may be appropriate to combine photoelectric light level detection, passive infrared levels, and ultrasonic motion detection. These sensors can be customized to increase artificial lighting levels based on motion detection or configured to perform different tasks based on the time of day. For periods of low occupation in the building, the sensors can be programmed to switch off or greatly reduce the lighting levels to address only specific tasks. Egress paths may also employ motion detection sensors to provide specific illumination and reduce light pollution and energy consumption.

Control management software is perhaps the most essential component of an ACS. This software will allow for management of multiple mechanical systems to maximize energy savings. Software can be set to specific levels allowing for individual occupants to have desktop control over task lighting in order to reduce the need for heavy overhead lighting, or temperature needs can be customized by zone and respond to heat gain. Air quality monitoring may be another valuable component of the system, and it can serve to manage airborne contaminant levels as well as work with ventilation systems to purify air.

Weather-related intelligence systems are a newly available feature of some ACS. This feature can connect to local weather and GIS data to predict outages of power, provide the framework for on-site renewable energy generation, or coordinate irrigation efforts with natural rainfall. In addition, the ACS may be linked to security systems and Emergency Management Systems (EMS) to alert direct-response authorities in case of emergency situations.

Consider the adaptability of the ACS, including what equipment can be controlled and how easy it is to reprogram the system. The building manager should be trained to monitor and use the ACS and a plan should be developed that outlines preventive maintenance. Test and calibrate the ACS at the intervals recommended by the manufacturer. Due to the complex nature of these systems, they should be regularly audited by a professional to optimize performance, assess efficiency, and ensure maximum occupant comfort levels are met. Professional auditing and commissioning should be a part of the long-term maintenance plan.

FURTHER RESOURCES

Websites:

1. *Lonix Intelligent Building Management System Specification Guide.* Lonix Ltd, n.d. Web. 30 June 2010. <http://www.lonix.com/specifications/IBMS_specification.pdf>.

2. *Installing, Retrofitting, or Upgrading Intelligent Building Systems.* Technologies for Facilities Management, BOMI International, 2009. Web. 07 July 2010. <http://www.fmlink.com/ ProfResources/HowTo/article.cgi?BOMI%20International:howto030509.html>.

Publications:

1. Ramsey, Charles, and John Hoke Jr.. *Architectural Graphic Standards, The American Institute of Architects. 11th ed.* Hoboken: John Wiley & Sons, Inc., 2007. Print.

2. Stein, Benjamin, John S. Reynolds, Walter T. Grondzik, and Alison G. Kwok. *Mechanical and Electrical Equipment for Buildings, 10th edition.* Hoboken: John Wiley & Sons, Inc., 2006. Print.

3. Bradshaw, Vaughn, PE. *The Building Environment: Active and Passive Control Systems. 3rd ed.* Hoboken: John Wiley & Sons, Inc., 2006. Print.

4. *Advanced Energy Design Guide for Small Office Buildings.* Washington: American Society of Heating, Refrigerating and Air-Conditioning Engineers, 2008. Print.

5. Bauman, Fred. *Giving Occupants What They Want: Guidelines for Implementing Personal Environmental Control in Your Building.* Berkley: Center for the Built Environment: University of California, Berkeley, 1999. Print.

6. Snowden, Dr. Jane L. *IBM Smarter Energy Management Systems for Intelligent Buildings.* New Town Heights: IBM T. J. Watson Research Center, 2009. Web. 07 July 2010. <http://www.citris-uc. org/files/Snowden%20IBM%20Research%20061009.pdf>.

7. *"Intelligent Buildings - A Holistic Perspective on Energy Management."* Electrical Review. London: St. John Patrick Publishers, LTD., 2009. <http://www.electricalreview.co.uk/features/118645/ Intelligent_buildings_-_A_holistic_perspective_on_energy_management.html>.

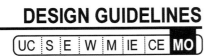
[MO.7] Hospitality Management Plan

SCOPE

EDUCATION | HOTELS | SPORTS

DESCRIPTION

Encourage projects to reduce waste associated with concessions and procure food from sustainable sources.

GUIDELINES

Concessions and food services can be a major component of education, hotels, and sports facilities, which provide food and beverage sales for building users including occupants, employees, visitors, event attendees, and support staff associated with events. The number of building users will determine the scale of the concessions provided by each building or facility, but the food and beverages consumed can generate a significant, negative environmental impact. A Hospitality Management Plan is necessary to outline steps and guidelines for the procurement, distribution, and waste collection associated with concessions and other hospitality-related services.

Food that is served in the building or facility should come from local, organic, and sustainable sources. Locally sourced food reduces the emissions associated with transporting food products and ensures food can be delivered and consumed soon after the food has been harvested or made. Food products that have to be transported great distances require additional preservatives or other measures to maintain freshness. Organic farming does not use synthetic products such as pesticides, chemical fertilizers, or genetically modified organisms. Procuring organic food products reduces the health risks on the consumers as well as the environmental impact of the farming system. Food procured from sustainable sources implies that the food production methods do not overly deplete natural resources. Sustainably sourced food has a lower environmental impact because the resources that are used to produce the food will renew themselves naturally. Develop a food procurement program that identifies the availability of local, organic, and sustainable food sources and how the project can maximize purchasing from these sources. Develop a delivery schedule to ensure food products can be used quickly to limit the potential for waste and spoilage.

Specify food service materials, including plates, cups, utensils, napkins, or beverage bottles, to reduce waste. When possible, specify reusable materials that can be returned, cleaned, and used multiple times. Provide appropriate service points to collect reusable food service materials to reduce any disposal of these types of items and educate building users on reuse programs. Specifying food service materials that are recyclable or compostable will also reduce the amount of waste generated through concessions. In addition to being compostable or recyclable, the food service items should contain a high percentage of recycled or

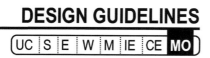
compostable materials as well. Avoid materials such as styrofoam that are not recyclable. Serve food with the minimum amount of packaging to reduce material consumption as well.

In addition to minimizing waste associated with food service items, limit the types and quantities of promotional materials that are distributed at the building or facility as well as at events. Advertisers and supporters should create promotional materials that do not require handing out large quantities of flyers or other promotional materials that will likely be discarded. Create a promotional display area or another type of commercial advertisement space for event sponsors rather than allowing mass distribution of promotional materials.

FURTHER RESOURCES

Websites:

1. *Sustainable Food Laboratory.* Sustainable Food Lab, 2011. Web. 12 April 2011. <*http://www. sustainablefoodlab.org*>.

2. Gustafson, Katherine. *What is Sustainable Food Anyway? Change.org, 05 October 2009. Web. 12 April 2011. <http://news.change.org/stories/what-is-sustainable-food-anyway*>.

3. *Sustainable Table.* Grace. Web. 12 April 2011. <*http://www.sustainabletable.org/issues/eatlocal/*>.

Publications:

1. Switzerland. *Green Goal: Legacy Report.* Switzerland: Federation Internationale de Football Association Organizing Committee, 2006. Web. 12 April 2011. <http://www.oeko.de/ oekodoc/292/2006-011-en.pdf>.

[MO.8] Sustainability Education & Awareness Plan

SCOPE

SPORTS

DESCRIPTION

Encourage projects to promote and educate employees, visitors, and local community members about sustainable initiatives and programs associated with the facility.

GUIDELINES

A crucial component of a sports facility is the presence of spectators and other visitors to the building. The sporting events themselves are an opportunity to showcase the athletic talent of the athletes, the culture of the host cities or countries, as well as the sustainable features of the building itself. Because sporting events can be inspiring to a wide audience, it is important for all of the components of a sporting event, including the athletes, the host city, and the building itself, to serve as a good role model for the audience. Also, as it can be a source of inspiration, design the sports facility to not only adhere to the highest standards of sustainability but to educate the building users on the sustainable features of the project.

In the planning phases, incorporate the input and feedback from a range of interested parties to ensure that the project is designed to meet the needs of the building users as well as the community. Conduct planning meetings with relevant stakeholders to determine the potential impacts of the project and how to mitigate them. Conduct outreach and education programs to ensure that the project's planning is transparent and understood by the community. Include a range of input from relevant stakeholders including local residents, organized community groups, environmental advocacy groups, business associations, and other interested parties.

In addition to designing the building to meet the highest standards of sustainability, create a program to promote and educate the building users about the sustainable initiatives and features. Also, by highlighting the sustainable features of the project, the sporting event can educate and inspire the building users. A program including signage or other graphic representations can educate users about the sustainable design features of the facility. Include descriptions of any design features and technical innovations, as well as how individual features work together to form a cohesive and sustainable project. Develop a Sustainability Education and Awareness Plan that reflects the comprehensive sustainability features of the project, with a focus on educating the building users.

In the design phase, plan for educational and promotional programs that focus on the sustainable features of the project. Plan educational programs and seminars that not only focus on the sustainable features of the project, but also on the principles of sustainability in general. Plan to host building tours for interested parties, including educational groups, community organizations, and the general public to further educate people on the sustainable design features and initiatives.

FURTHER RESOURCES

Websites:

1. *The US Partnership For Education For Sustainable Development.* Washington: US Partnership for Education For Sustainable Development. Web. 12 April 2011. < http://www.uspartnership.org>.

2. *Envirosax: Sustainability Education Resources.* San Diego: Envirosax LLC, 2004. Web. 12 April 2011. <http://www.envirosax.com/education_resources>.

3. *Second Nature: Education for Sustainability.* Boston: Second Nature, Inc., 2011. Web. 12 April 2011. < http://www.secondnature.org/>.

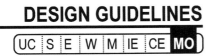
[MO.9] Building Legacy

SCOPE

SPORTS

DESCRIPTION

Encourage projects to plan for the continued operation of a facility following the initial events for which it was designed.

GUIDELINES

Sports facilities that are designed and built to host a major sporting event have to address how the facility will be used following the initial event. While the design and function of the building during the initial event should meet the necessary capacity and technical requirements of the initial major sporting event, the building must have a lasting and positive legacy on the environment and the community. By planning for the legacy of the project in the design phase, the project will be able to plan for sustained and continued operations for the remainder of the life span of the building.

Although a sports facility is designed for a particular sport or function, the Legacy Plan will address the additional functions and events the building can host. Determine what types of other competitions and sporting events the building can host. Also, the building can host the same types of sports from different leagues or levels of competition. In addition to hosting other sporting events, determine if the project can host different functions such as concerts, shows, or conventions. Design components, such as competition venues and spectator seating, to allow for different configurations and setups to accommodate a wide range of activities in the legacy phase.

The Legacy Plan will address how the building supports the local community in terms of access to the facility as well as community-building programs that are hosted. To support the local community, allow local residents access to the facility. In particular, provide reduced costs and easier access for low-income and under-privileged residents. Additionally, a sports facility can host events and programs to engage the local community. These programs can focus on sports and athletic training or education, as well as other community building speakers and programs. A community-focused education program will engage the local community and provide a positive legacy.

Furthermore, a sports facility has to plan for the successful financial operation in the legacy phase. The Legacy Plan must outline the revenue sources the project will rely on to operate in a financially sustainable way. These revenue streams can come from the additional events and programs the building hosts. For this reason, it is important to consider the types of events the building will host in the legacy phase and the potential revenues these events will generate. Developing a viable financial plan in the design phase is important to ensure the building can be maintained and operated for the lifespan of the building.

The building can conduct renovations to reduce capacity or increase the flexibility of the facility for the continued operation of the building in the legacy phase. Reducing capacity will eliminate extra spectator seating, parking spaces, and other facilities that would not be required. Reducing capacity of the facility has the potential to reduce energy, water, and material consumption. Facilities that undergo renovations will be rated through the GSAS Operations, but it is important to plan for the sustainable renovations that will take place in the design phase. Reuse and recycle building components that are removed and procure new materials from sustainable sources. Specify materials with high recycled contents and low VOC contents. Plan renovations to improve the efficiency of the building in the legacy phase and carry out the renovations in a sustainable way to minimize the environmental impact.

FURTHER RESOURCES

Websites:

1. *2012 London Olympics: Olympic Park Legacy.* London: 2012 Olympic Organizing Committee, 2011. Web. 12 April 2011. <http://www.london2012.com/about-us/the-people-delivering-the-games/stakeholders/olympic-park-legacy-company.php>.

2. *Olympic Park Legacy Company.* London: Olympic Park Legacy Company, 2011. Web. 12 April 2011. <http://www.legacycompany.co.uk/>.

3. *Sport England.* London: Sport England, 2011. Web. 12 April 2011. <http://www.sportengland.org/>.

APPENDIX I: DESIGN & CONSTRUCTION OF HVAC SYSTEMS

This Appendix is intended to provide system information, recommendations of good practice, project considerations, and discussion of options for the design of the HVAC system. Additionally, this section discusses options, considerations, benefits, and concerns with each part of the HVAC system and represents attempts to standardize the approach to design and specification of the HVAC system. The following describes the steps involved in designing a HVAC system.

Defining the Scope/Goal of the Design

Defining the scope/goal of the work begins with the engineer collecting the following information:

- How large is the building and what kind of needs are being satisfied in the building design?
- Who are the ultimate users of the building?
- What are the available cost effective and reasonable options of utility that can be used to run the building?
- What is the expertise level of the maintenance staff that will take over control and maintenance of the building after occupation?

The engineer should use the collected information to limit the choice of system design options after discussing and receiving confirmation from the owner.

Designing the HVAC System

Steps in creating the Contract Document:

Once the project scope is well understood, the engineer should start designing a system that can meet the design goals. Elements to be considered are:

- Local climate (outdoor air temperature, outdoor humidity, wind speed, sky clearance, etc.)
- Building characteristics (size, orientation, material, etc.)
- Indoor temperature and humidity controls (level of comfort for occupants, specific purpose spaces such as data center, etc.)
- Air movement, change and quality
- Air and water velocity
- Running a load calculation for specifying cooling and heating capacity requirements (Usually maximum cooling load can occur at peak wet bulb conditions when outside air demand is high; therefore, for sizing equipment peak total load, both sensible and latent, climatic conditions should be considered. When designing the central plant the designer should use methods such as life cycle analysis as a basis for optimum design.)
- Space requirements (It is important to leave enough space for servicing and cleaning the equipment. Consider the location of the outside air intakes relative to the exhaust outlets, and the plant and equipment room location in relation to the spaces that they serve, sufficient horizontal and vertical spaces, etc.)
- Noise control considerations (Design to minimize sources of noise such as direct transmission of mechanical equipment room noise to adjacent spaces, duct-borne noise generated by fans and ducts, duct rumble, etc.)

- Maintenance and operating cost
- Running an energy analysis to verify energy savings
- Inspect the effect of certain choices on the local Global Sustainability Assessment System (GSAS)

The engineer should also consider the design system limitations such as capacity availability for specific equipment, and available space for the equipment.

After specifying the required cooling and heating for the building through a load calculation process with one of the available load calculation tools, such as Trane or HCCV, the HVAC engineer shall select the appropriate equipment for the design.

The most important equipment and accessories-related choices are the following:

a. Refrigeration Equipment

- Chillers
- Cooling Towers
- Chilled and Condenser Water Pumps

b. Heating Equipment

- Boilers
- Unit Heaters
- Hot Water Pumps

c. Air Delivery Equipment

- Air Terminal Units
- Duct Insulations
- Above Ceiling and Under Floor Plenums

d. Chilled and Hot Water Pipe Distribution

- Pipe Systems
- Pipe Insulation

e. Space Requirement

- Security
- Equipment Rooms
- Horizontal Distribution
- Vertical Chases
- Equipment Access
- Rooftop Units/Fans

f. Automatic Control Systems

g. Maintenance Management Systems

h. Building Commissioning

The engineer should select one or a combination of the following systems based on the applicability to the project:

1. De-central Cooling & Heating Systems

Heating and cooling for the building is produced in different locations including inside the occupied zones.

Advantages:

One of the best advantages of de-central cooling and heating systems is the fact that each space can have independent cooling and heating at all times. The system is also appealing because usually one single manufacturer constructs the equipment and therefore, the efficiency of the system is not dependent on other equipment. Installation of the system is fairly simple and if there is a malfunction in the system, only a small part of the building—a single zone—is affected.

These types of systems are appropriate due to their availability, and simple operation eliminates the need for expert operators. They are energy efficient due to the capability of turning the zone equipment off when there are no occupants. Finally, there is no need for large mechanical rooms, initial costs are lower, and the utility uses can be measured in each zone (for multi-tenant assemblies).

Disadvantages:

The main disadvantage of de-central systems is the fact that their performance options are limited due to standard sizes that are not easily adjustable. Another drawback is that applying diversity is not possible; therefore, the total installed load and the utility consumption is larger than what is actually required. A high level of control is not possible, and use of an economizer is not an option. Another problem is that the equipment and its generated noises are closer to occupied spaces.

Finally, in these systems, a condensate drain piping throughout the building is required and maintenance is more difficult due to multiple installations of equipment in the building.

2. Central Cooling & Heating Systems

Central cooling and heating systems are typically systems that produce required cooling and heating (except perhaps some supplemental heating) in a central mechanical room.

Advantages:

The direct advantage is the fact that cooling and heating is available at all times. Central systems require less service space and also less maintenance due to a smaller quantity of equipment. Other advantages of central systems are diversity and redundancy availability, and at larger capacities it translates to better part load efficiency.

DESIGN GUIDELINES

Disadvantages:

One of the disadvantages of central systems is that equipment is not ready off the shelf, and may take a long time to be delivered. Another disadvantage of central systems is that they are more complicated and therefore, need more experienced operators and maintenance people.

Finally, other negative characteristics of the central systems are the fact that a relatively large mechanical room with extended height is needed, more complicated controls are required, and a large piping distribution system will be required.

3. All Air Systems

All air systems provide the sensible and latent cooling capacity through cold air supplied to the conditioned space, and no additional cooling will be provided in the zone. Typical all air systems are Constant Volume, single duct, terminal reheat; Constant Volume, double duct; Multi-zone; VAV single duct and VAV dual duct.

Advantages:

One of the main advantages of this type of system is that the maintenance of the equipment will be done in an area removed from the occupied zones, and any impacts from vibration and noise generating equipment can be controlled well.

There is the potential of using an economizer cycle, and systems generally have a higher degree of flexibility for optimum air distribution, draft control, and adaptability to varying local requirements. Other benefits are smaller units due to the possibility of diversity, better control availability, and centralized condensate drain piping in mechanical rooms.

Disadvantages:

The main disadvantage of all air systems is that the existence of larger ducts causes less floor-to-ceiling height availability, and a larger building footprint may be required to accommodate the mechanical rooms and vertical chases. Other disadvantages include complicated air balancing procedures, and if one of the major pieces of equipment goes out of service, most of the spaces will lose cooling or heating power.

4. In-Room Systems

In-room systems condition spaces by providing air and water supplies to terminal units in zones. Typical elements in this type of system include primary air with an induction unit; primary air with a fan coil unit; water source heat pump; fan coil unit; perimeter radiation; and radiant panels.

Advantages:

Benefits of in-room systems are the availability of individual room control, less space requirements due to smaller distribution ducts, and the ability to continue operations even if a unit is down.

Disadvantages:

A disadvantage is that more controls are needed, even though they are less complicated. Low primary chilled-water temperature and/or deep chilled-water coils are needed to control space humidity adequately. The system is not appropriate for areas that need large quantities of outside air or exhaust, and a large number of condensate drain pipes may be needed.

5. District Cooling/Heating

District cooling systems are utilized best in areas with high building density. They require a large investment initially, but cause less pollution than individual systems due to larger and higher quality equipment and higher energy efficiency. They also require less maintenance.

District cooling is one of the most efficient methods of providing cooling to individual buildings. Both scale effects as well as advanced cooling plant technologies guarantee that this is the most appropriate solution. The government of Qatar provides this facility for most of the country and new buildings should be planned in locations where cooling plants keep pace with changing demands.

APPENDIX II: ENVIRONMENTAL CONDITIONS OF QATAR

Qatar Hourly Temperature of Each Month

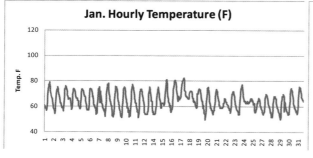

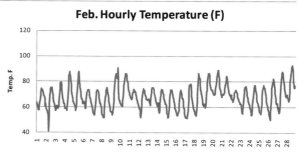

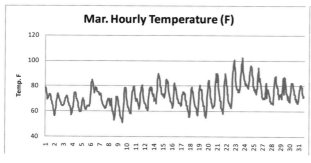

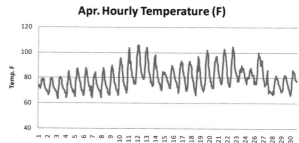

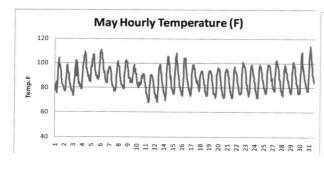

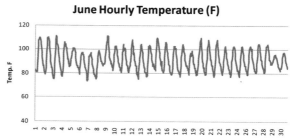

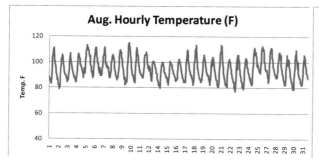

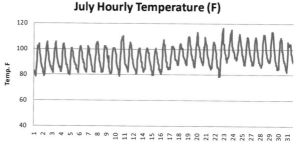

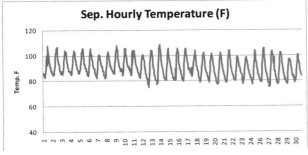

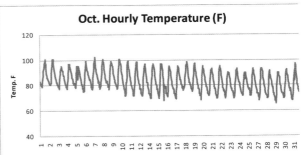

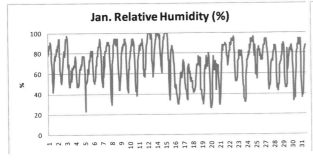

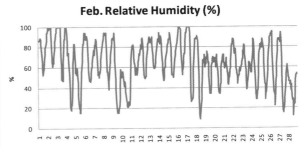

Qatar Hourly Relative Humidity of Each Month

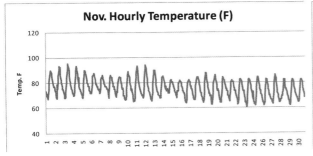

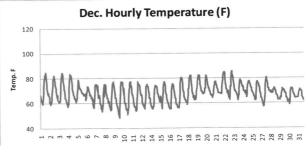

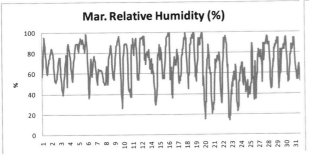

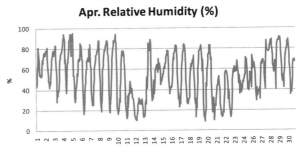

May Relative Humidity (%)

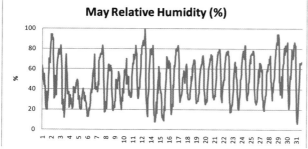

June Relative Humidity (%)

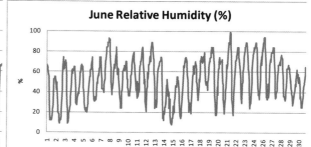

July Relative Humidity (%)

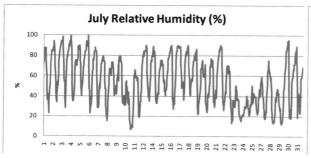

Aug. Relative Humidity (%)

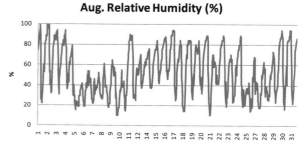

Sep. Relative Humidity (%)

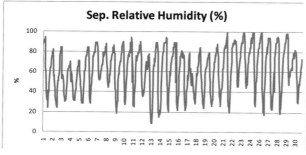

Oct. Relative Humidity (%)

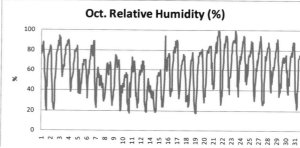

Nov. Relative Humidity (%)

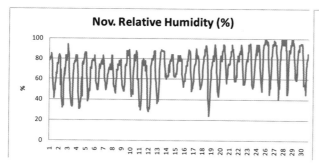

Dec. Relative Humidity (%)

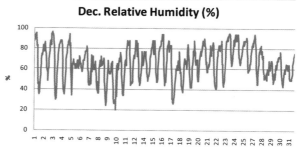

Made in the USA
Charleston, SC
29 November 2013